Terroir and Merroir

Environmental factors impacting taste
Terroir is land. Merroir is water.

If terroir and merroir set the place for culture,
and seeds are nature's intersection for folklife
and foodway, then this is the Jersey tomato
version of that.

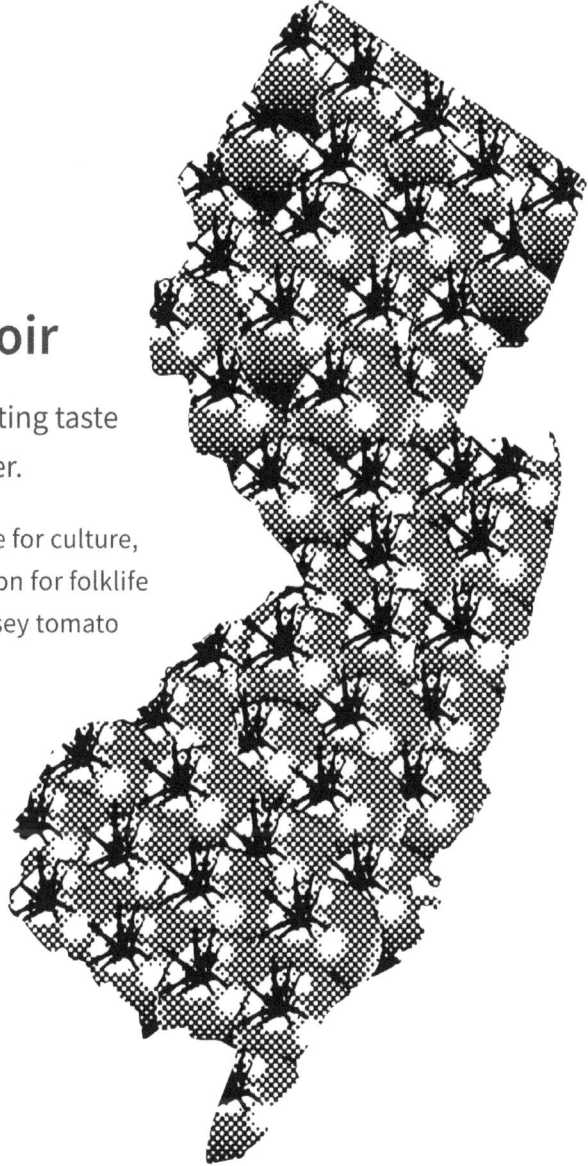

Iconic Jersey Tomatoes An American Folklife and Foodway

By Jeff Quattrone

New Jersey is the traditional territory of the Nanticoke
and Lenape people, who call it "Lenapehoking."

This book recognizes that, and the stolen land,
trauma, and genocide committed against them here
in New Jersey and against all Indigenous people
in the United States and around the world.
Genocide should never happen.

This book recognizes slavery in New Jersey
and the human rights abuse that happened here
and in the United States, and around the world.
Enslaving humans should never happen.

This book recognizes the origin of wild tomatoes
in the Andes of western South America,
eventually domesticated by Indigenous communities
before the colonizers arrived. They brought them to Europe,
who brought them back to America when they colonized here.
Acknowledging the origin of food should always happen.

Table of contents

Foreword

SEED LIBRARIES

Seed and knowledge, nourishment for your soul.

Seeds are the genetic storehouse of the plants they produce.

Libraries are the storehouses of knowledge produced by society.

I didn't set out to live and write the story I'm about to tell. At some point I decided to look at life as a story I lived every day. I wanted to recapture the wonder and amazement I had as a kid between the covers of the books on my bookshelf at home or at my local library.

Life's a story. It's a collection of personal, social, and often unexpected experiences. In my story, the most unexpected and inspirational experiences revolve around seeds, in all their complexity. Seed activism changed my life, and in turn I was able to affect permanent social change with free seed programs in South Jersey, where I live. I did this with my Library Seed Bank project, which brought the concept of seed libraries to South Jersey. From there I advised others throughout New Jersey, across America, and in Calabria, Italy, where my father's family migrated from and where I'm registered as a citizen.

You see, seeds don't need us, but we need them. Nature created seeds with the sole purpose of germination so they could produce more seeds to keep evolving year after year. While they're doing that, they provide sustenance for us on Earth. When humans started domesticating wild plants for agriculture, seeds became intertwined with all the good and bad that is human nature.

It is through seeds that I found my story. Did I find seeds, or did they find me? In the end, it doesn't matter.

The story of seeds belongs to the people who stewarded them before me. Archives contain the past in many forms. Seeds that are saved are an archive, an intersection of all who have brought the seed forward. They offer the opportunity to connect to the past and correct records by telling the complete story of slavery and genocide connected to seeds, so we can move forward together knowing that seeds sustain us. By saving and sharing seeds we can coexist with a shared purpose.

It's a Jersey Thing

All localities have their cultures. The difference here is our attitude. For better or worse, those of us who are from Jersey, are passionate, bold, and unapologetic. It stems from deep pride in where we live and where we are from.

While primarily the son of Joseph and Margaret Quattrone, I'm also an adopted son of Jersey. I moved here when I was five, and have been rewarded with this transcendental story of seeds, local culture, and tomato breeding. It's such an honor to bring forward the seed and tomato breeding history, and share the lives and passion of the folks who contributed to a seed- and food-secure circular local economy. Generations of individuals, families, and communities all connected through the seeds, tomatoes, and commerce that inspired innovation in tomato breeding and glass and canning technology. It's no wonder the Jersey tomato is so iconic to us and our culture.

There are a lot of jokes about Jersey and how bad it is. Some of them are true. At the same time, we know that nowhere is perfect, and we are too busy enjoying all that our terroir has to offer to dwell on that.

We're a peninsula that starts with a narrow river, the Delaware. That's our border to the west. The Delaware flows down and around the southwest portion, creating a rambling western edge. Once around the curve that is Salem County, the Delaware River flows into the incredible and dynamic Delaware Bay and its stunning coastline. The rustling of wind and the sharp cries of the shore birds, the aroma of salt air make you feel the magnificence of our existence. You can taste the Bay in the air.

The Bay is an ecologically important site for horseshoe crabs and shore birds. It meets up with the Atlantic Ocean at Cape May Point, the southernmost point of the marvelous Jersey Shore. It is a one-hundred-and-twenty-mile mix of barrier islands and shoreline with eclectic communities, beautiful beaches, and an unfortunate inclusion in a really bad reality series.

In between the river and ocean, we have fields of tomatoes, peppers, eggplants, and corn. There are orchards of stone fruit, apples, and pears. We have dairy farms, along with pastures of cattle, pigs, and chickens, lots of horse farms, and the crown jewel of the state, the Jersey Pines, home to over twenty varieties of native berries. And there are the woodlands that offer an abundance of tastes and textures for foraging.

Our waters are fertile with shad, oysters, scallops, crabs, clams, and a variety of fin fish, including flounder. Catching your first flounder is a rite of fishing passage; gotta love a flat fish.

Within New Jersey, there are three distinct areas: North, Central, and South. You will find debate about whether Central Jersey exists, but it does. One simple way to discover what area of New Jersey someone is from is by asking them if a local breakfast meat is Pork Roll or Taylor Ham. North Jersey calls it Taylor Ham, South Jersey calls it Pork Roll, and Central Jersey for the most part calls it Pork Roll. No matter what it's called, it looks like salami and tastes similar to bologna. It's greasy and delicious on a breakfast sandwich or on a burger.

But when it comes to something that the whole state claims, we unite under the Jersey banner.

Like the Jersey tomato. A true American folklife and foodway. It has a loyal and passionate regional following. Iconic if you will. I'm here to make a case for it to be a national icon, and I have the receipts.

It's part of my Jersey trifecta: the Jersey Shore, Jersey Girl, and the Jersey Tomato. That sums up my life in Jersey from the time my family moved from South Philly to the newly minted suburbs in Turnersville, NJ, to me writing this book.

I have to give a shout-out to Bruce Springsteen, who influenced me to become an artist. I didn't know what I wanted to study when I graduated high school. I went to community colleges to try to figure it out. Not satisfied with my original goal of being the best middle manager that the world would ever see, I felt something more was out there for me. This escalated quickly when a professor assigned us *Dress for Success* by John T. Mollory for a salesmanship course.

The campus bookstore didn't have it in stock, so I went to the mall. When I walked into the bookstore, there was a display of *Born to Run: The Bruce Springsteen Story* by Dave Marsh. Having danced the night away with Bruce, the Big Man, and the E Street B Band many, many times at the Spectrum in South Philly, I bought that book instead. I mean wouldn't you? If not, you're likely not from Jersey.

The next day I cut my salesmanship class and started reading Bruce's story. I sat at the community college in the student center for most of the day doing that.

I realized that if this guy from Jersey could go and do what he really wanted to do, then so could I.

I went to talk to a counselor, and said I wanted to get into radio and TV production. They suggested that I check out what was then Stockton State College. (It's now Stockton University.) Stockton was having an open house soon after that meeting, and I went. I fell in love with the campus. I applied. I got accepted, my courses transferred, and I was all set. Then, three weeks before my first semester began, I got a letter saying that budget cuts killed my program, and I was rolled into a visual arts program.

I had loan papers signed. I had already bought a 35 mm film camera that was required for the major I applied for. I was commuting my first year, so I figured why not see what it was about. Once I started studying art, I discovered this beautiful and ancient visual language. I decided to stay in the program and become an artist.

This is one example of how deeply rooted my life is in Jersey culture. A significant Jersey artist's everyman persona influenced me to pursue my passion. I didn't realize that art would become my passion and my life. I discovered it at a state college in New Jersey, at a school with the tag line *Nestled in the Pines*." The Pines being the Jersey Pines, a unique ecological area that is home to fauna and flora that exists there and nowhere else, and some of the longest generational inhabitants in the state's history.

It wasn't until I started writing this book that I realized the significance of this connection between studying art and Jersey culture.

This art education let me discover my creative process, which is working in chaos. That's where the magic is for me, and for the story I'm telling you. I didn't plan on being an artist. The huge curve ball of having my major changed three weeks before I started isn't something you plan for when deciding what you want to do for the rest of your life at eighteen or twenty.

I also didn't plan on getting into seed work. Just like the fated fork in the road that started my artistic path, when I started with seed work, I had no idea of all the remarkable, layered, and complex stories associated with seeds.

My work with seeds came about after a fated trip to Slow Food International's Terra Madre food festival in 2012. Slow Food is an international organization dedicated to good, clean, fair food for all. I had been aware of Slow Food since the late 1990s, and it was through Slow Food's Ark of Taste project that I discovered food biodiversity, and the threat of local food becoming extinct. The Ark of Taste is a living catalog of local food biodiversity that's at risk of extinction due to climate change, habitat destruction, and consolidation in the food industry.

I applied to be a Slow Food USA delegate but was not chosen. I was offered an opportunity for a press pass. I had a blog at the time about the loss of food biodiversity, but couldn't find a way to position my voice in a crowded field of creators and activists working on this issue.

Those four short days turned into an extended stay thanks to Superstorm Sandy. I was stranded for a week in Rome. During the festival, my mind was blown by the scale of activism on display. I was also inadvertently blessed by the Pope on All Saints' Day.

I'm now an atheist. I was raised a Catholic, so when I found myself walking towards the Vatican approaching noon, I realized from my youth what was about to happen. The Pope came to the window and called out the groups of pilgrims assembled below him, who cheered when their group name was called. And I was standing there because nature caused a huge ruckus in the New Jersey/New York area.

When I got back, I was committed to being a food system activist, but still didn't know how. In January 2013, I saw my first article about seed libraries, so I committed to becoming a seed activist. If you don't save seeds and keep them viable, they go extinct. That's the core of food seed activism for me.

After that I became a co-chair for Slow Food USA's Northeast Ark of Taste Committee, and suggested seed programing ideas for the 2018 Terra Madre festival. I was invited to join an international community seed saving panel at the 2018 festival. In 2021, I co-organized the first Slow Food Seed Summit as part of the 2020 six-month virtual Terra Madre, due to COVID.

I often say that I am an artist by accident and an activist by choice. I didn't choose to be an artist when I got out of high school, but by circumstance or providence I ended up becoming one. I was taught about reading symbols and the communication that comes with them. A connection to the intuitive. Out of all that great stuff that happens when you're intentionally living the story of your life, foreshadowing is one of the best. You only see it looking backwards, but it's a key storytelling element that has monumental impact.

My family and I shopped at Ledden & Sons in Sewell, New Jersey. When we moved from South Philly to South Jersey in the mid '60s, there were no big box retailers. We were on the edge of tract housing taking over farmland. Within a year after we moved, we got a shopping center with a Grants department store, Thrift Drug, and a Pathmark supermarket, but not much in the way of gardening supplies.

We moved in October, so setting up a garden in the backyard had to wait until the spring of the following year. That's when we started going to Ledden. It was a farming supply business, and the Ledden family were also farmers.

Ledden had a seed room. As soon as we parked, I would bolt there. I was drawn by the wonder of it. There were drawers filled with seed packages, barrels filled with onion sets and seed potatoes during planting season, scales, farmers in overalls, and this musty earthy aroma. A squeaky wooden floor. The set design was superb to inspire wonder and fascination in a curious kid. I look back on this now, and I realize that sensory experience planted the seed of where my life would up end up decades later.

Ledden also carried seeds of the local tomato breeding that was happening at that time. In the '60s when we would go, the tomato breeding was winding down from its heyday, but there were still local farmers who provided local farming supply companies with seeds. Ledden was a generational farming family, with a farming supply company. They had their own variety of celery, and they were the largest supplier of plants in the state according to their 1936 catalog. Leddens is another layer of Jersey culture that intersects with tomato breeding in my story.

Growing tomatoes in South Jersey is a summer ritual, as it is in many places in the world. Thanks to my Dad and his desire to turn a big portion of our backyard into a garden, I was initiated into this ritual with lessons about organic gardening. My dad was a big fan of Rodale's, a leader in organic farming and education. We followed their advice, and he knew an organic farmer. My parents were first-generation Italians, and my grandparents were part of the large Italian migration to Philly in the early 1900s. Both sides of our family knew farmers through friends and family. So there was a lot of free manure for our garden, and random bushels of tomatoes, peppers, or eggplants at any given time.

My dad was always on the quest for knowledge. He was the oldest of his siblings, and like many of the oldest children in a family at that time, he dropped out of high school to help support his family. You would mention something in passing and a day or two later, out of his archives, he would have some information for you. He taught my bothers, Joe and Mark, and me about organic growing, and tomatoes were a big part of his garden plan. This is where my passion for tomatoes started.

Growing tomatoes was only half of the equation in our household. Canning them was the other half. The month of August was all about preserving our harvest, and the harvest that came our way with the random baskets of produce from our family friends' farms.

My mom loved to cook, so for her, canning the local flavors for her family for the winter was something she relished. We had the *"shelves under the stairs"* in our basement. There were three shelves under the basement stairs, and they would be almost filled with cases of tomatoes, pickled peppers, pickled eggplant, zucchini relish, peaches, and orange marmalade. Granted the oranges weren't local, but my mom loved orange marmalade, so we all got the benefit of that.

I also learned about growing flowers from my mom and her brother, Uncle Frank. I didn't realize at the time how important the flowers were for pollination. I understood how pollination worked, and back then insects weren't under the pressure that they are today. My uncle would come to visit, and he, my mom, and I would take a ride to the nursery right by Ledden to buy the stock he needed for the gardens at Stella Maris Church in South Philly.

Sometimes I would take a ride with him back to Stella Maris and work in the gardens. My mom would come by and scoop me up and we'd have a quick visit with my grandmother and aunts who lived in South Philly.

I was a lucky kid. I was learning lifelong lessons about seed and food sovereignty at a time when teaching it wasn't necessary. It was everyday life. It started in the seed room at Ledden, carried through to growing both food and flowers in the garden, and finished with canning in the kitchen. It all came together as seed and food sovereignty, the Jersey way.

The Terroir, People
and Culture of Jersey

Jersey history, like all history, is a collection of stories that reflect the spectrum of humanity. The eccentricity, the trauma, the triumph, and the tragedy that are our collective bond. It's the local terroir and history that create the culture that gives each location its place and unique identity.

The Jersey tomato is an intersection of local pride and culture. It developed from our terroir. It doesn't get much more local than that.

Since seeds are the lens of this story, it's through them that this story was revealed to me. If these seeds went dormant forever, the passion and history of those who created them would also be dormant. Because of the foresight of companies and educational and research programs donating seeds to the USDA, and individual seed savers, this vibrant connection to history, community, and the terroir of our peninsula is sustained. We have universal access to respect, revive, and reinvent the seed and food sovereignty around the Jersey tomato.

Jersey's terroir is conducive to growing crops. It was the intentional focus of breeding great-tasting tomatoes that yielded well and were disease-resistant that made the Jersey tomato what it is. The folks breeding them had a unique terroir to experimentally breed and farm here. And they hit the jackpot.

The uniqueness comes from our varied terrain and climate. The Skylands in the north. The Delaware River, other rivers, creeks, ponds, lakes. Their associated wetlands throughout and along the Delaware Bay. The Jersey Shore and Atlantic Ocean. The astonishingly bio diverse Jersey Pines. A true four-season climate.

That terroir is present not just in tomatoes, but corn, berries of all kinds, straw-, black-, blue-, and cran-, and the peaches, nectarines, apples, plums, and pawpaws.

There's the sea mineral taste from the Delaware, Barnegat, and Great Egg Harbor bays. Clams, crabs, scallops, and oysters. Oysters have a proud history here in Port Norris and Tuckerton. The Maxwell family in Port Norris farmed oysters in the brackish waters of the Mullica River in the rare and bio diverse Jersey Pines. Fin fish from the freshwater, the brackish water in the bay and in our rivers, and from the ocean.

Our terroir feeds our bodies, stimulates our senses with the terrain, and fuels our passion. And with the intentional breeding of Jersey tomatoes, for flavor, we're damn proud.

It's the peninsula that separates us from the rest and that allows for the development of our distinct Jersey flavor in cuisine, history, and culture. New Jersey is one of four states in the U.S. that are peninsulas (the others are Alaska, Michigan, and Florida). The peninsula forms natural borders, so after the Revolutionary War, Jersey had no further western land to colonize. With New York to the north, and Philadelphia to the south, we never had a major port or large city. A lot of our commerce was through those ports, which built more generational wealth there than here. Self-sufficiency influences the independent and proud Jersey culture.

New Jersey is also the only peninsula state that is bound by freshwater, brackish water, and ocean water. Pools of water in the form of creeks, streams, lakes, and rivers traverse our interior soil. Tidal actions mix in places with some of the best soil in the country for growing crops, and in other places supporting the unique bio diverse Pines. We have three growing zones: 7, 7a, and 7b.

Jersey soil is used on all new baseballs in the Major League. This magic mud is used to roughen the smooth texture so the pitchers can grip the ball tighter. Russell Aubrey "Lena" Blackburne, a coach for the old Philadelphia Athletics, created the product after an umpire complained to him about spit and sawdust that was being used at the time to roughen up new baseball.

Blackburn lived in Jersey, and when he got home one day, he checked out the mud in his fishing creek. He tried it on a baseball, and it created a surface a pitcher could grip without making the ball soggy. It dried to a powder that could be removed.

The exact location of the creek is a secret, but it is around Pennsauken on the Rancocas Creek. There, tidal action washes up "sugar sand," the local name for Pine Barren's soil. That soil, and organic matter from the Pines, wash up to the Pennsauken and Rancocas Creek, tributaries of the Delaware. The Pines are east of these creeks. Tidal actions traverse our soil and unite two of the three different types of water that contribute to our peninsula terroir, creating the Magic Rubbing Mud cauldron.

We have four soil types here and five soil zones according to a Soil Zone map from *Agriculture in New Jersey*, a 1941 New Jersey Department of Agriculture publication. You can see them on the replica of the map on the next page. Zones 4 and 5 comprise about 60% of the state with two different soils. This area is also the portion of Jersey on the Coastal Plain.

New Jersey Soil Zones

approx. location
of source for
Magic Mud

Zone 4 comprises approximately the western third of that area and is referred to as the Heavy Coastal Plain belt. It's some of the most fertile soil in the country for truck-farming crops, including tomatoes. Zone 5 is referred to as the Sandy Coastal Plain Belt, aka sugar sand by the locals. It's suited for market crops, small fruits, and blueberries and cranberries. It's also the soil of the awe-inspiring Jersey Pines, also known as the Pine Barrens. Early colonizers attached the word "Barren" to this incredibly diverse pine forest because they couldn't farm their crops there.

Magic Rubbing Mud is found in the Heavy Coastal plain, with the sugar sand from the Sandy Coastal plain washing upstream with organic matter from the Pines. This organic matter is found nowhere else.

From east to west, New Jersey is about 77 miles wide from river to ocean, and approximately 134 miles north to south. There are three growing zones within that small area. Three different marine bodies influence a relatively small area, with pools of water running throughout with two distinct soils: one where blueberries and cranberries are native fruits, and one that has some of the best soil in the country for vegetable crops. Until climate change, it was a relatively stable climate with four distinct seasons.

I envy the tomato breeders who had this fantastic terroir to work with when the climate was stable. But I will adapt to the new climate, and carry this proud tomato breeding legacy forward.

When you have a food that was intentionally bred to thrive in a particular terroir, it is important to build commerce and community. Fortunately, from the start, taste was a key consideration in breeding Jersey tomatoes.

Once ketchup started to become a consumer product, after the Civil War, tomato farming ramped up in Jersey. Jersey was part of a ketchup production area from upstate New York to South Jersey, over to Indiana, and back up through Michigan.

A local circular economic and social structure was built around the Jersey tomato. Tomato breeders, like those who were working with Campbell's on developing new varieties, farmers, canners, and seed companies secured food for the community, hence seed and food sovereignty. This wove a social fabric that connected all, because the tomato is a food that was accessible to all. Scaling up the production of tomatoes through innovative tomato breeding with strong plants that thrived locally and produced abundantly encouraged innovation in the production of cans and bottles for the products.

The intersection of these industries, and the people who worked in them, allowed locals to put down roots. They had families and built communities with rituals for generations. Campbell's gave out plants to their contract farmers and their employees. Their tomatoes were grown in the backyards of the folks who worked for them. These tomatoes were shared with family and neighbors. The breeding innovation was shared by all, whether they knew it or not.

While the tomatoes were bred to thrive broadly, the terroir here played an essential role in developing them. Cans and bottles can be manufactured in many places, but tomatoes responded so well to our soil and climate. Jersey has good sand for glass too, but there was no intentional process to create a Jersey glass. There was a lot of intention, though with Jersey tomatoes.

A lot of canning was done on farms, or within an approximately 50-mile radius in Campbell's case. It was all local. There were multiple companies with seeds, experimental farms, and canning facilities. And according to Mary B. Sim, writing in *History of Commercial Canning in New Jersey*, the first recorded reference to tomato farming in New Jersey occurred in 1812 in Bridgeton, a city in Cumberland County. According to Sim, there were no other records of tomatoes being grown in any northern states.

The Jersey tomato is an intersection of Jersey culture, tomato seed breeding, and local seed and food history. The seeds are the story. Without them, there would be no tomatoes. The new seeds created by the tomato breeders carried on the living history of innovation, passion, and dedication of New Jersey tomato breeders who pushed taste, production, and disease resistance to new levels.

This is an example of the complex and layered relationship between humans and seeds that began when humans started to domesticate wild plants, leading to what we know today as agriculture. The tomato seeds tell a local story with their history and genetics.

And here in Jersey, we are passionate about them. While we can't agree whether a local breakfast meat is a Pork Roll or Taylor Ham (it's Pork Roll), we all agree that our tomatoes are the best. Culture, community, and history all come together through the seeds that were bred for a prized culinary experience and built communities for generations.

In these communities, the passions and life's work of the people involved at all levels of growing and production compose the history and culture. For those not directly touched by the tomato industry, the locally-bred tomatoes were on their tables for celebrations, memorials, and likely in canned jars in their pantry.

One day I was in Camden, NJ. Campbell's headquarters has always been in Camden, and for a long time, so was a lot of their soup production. I had just picked up the first large harvest of the Garden State, a Campbell's-bred tomato and our state's official nickname, from the Center for Environmental Transformation (CfET), an urban farming and environmental organization in the South Camden Waterfront. CfET is a wonderful community-based organization that brings fresh vegetables to their neighborhood. They hire local youth to learn about growing food and the nutrition of what they grow. They lead in environmental activism. I've worked with them on a few projects related to Campbell's-bred tomatoes, and I'm proud to partner with them. I would like to thank Teresa Neida, Dean Buttacavoli, and Jon Compton, who I met through CfET for their collaboration.

After picking up the harvest, I drove over to the Camden County Historical Society to research Campbell's. It was about a mile away in a different neighborhood. As I was getting a few tomatoes to bring to the librarian I was working with, I looked down at one tomato in my hand, and I realized "There it is."

This was the first time that I saw the Jersey tomato as an intersection of all things Jersey. I knew it conceptually, but it became real by scheduling a convenient pickup of a harvest of tomatoes bred by Campbell's, from seeds donated to the USDA Germplasm Bank by Campbell's. This tomato was grown in the soil of the historic home city of Campbell's, about twelve miles away in the same county as their breeding program, and I was about to share this tomato with a person helping me research Campbell's history for this story I'm telling you now.

What I scheduled as a convenience turned into an unexpected defining moment of this project. And if I didn't have the seeds, none of that would've happened.

Seed history, plant breeding, slavery, canning, backyard gardening celebrations, memorials, and livelihoods. It's all here.

Edgar Hurff used the tag line *We're the Jersey tomato people* for his canning company. Hurff had a seed company, too. The P.J. Ritter Co., a specialty food company in Bridgeton, also had a seed division, the Ritter Seed Company, and they had an experimental farm where they bred their own tomatoes and peppers. Then there were the seed-only companies, and the farmers doing field trials for Campbell's. In one experimental tomato farmer's case, he was breeding his own varieties of tomatoes, which he sold seeds from his farm. He sold them through a local farming supply company with a seed room, and through a seed company where his tomato had a national presence. The seed complexity of these local companies generated community participation; they represent the macro and micro social relationships engaged during the heyday of Jersey tomato breeding.

There's Warhol and his Soup Can Series. The tomato soup can is arguably the most popular from that series. So much so that *Esquire* magazine had Warhol being swept up in a whirlpool of tomato soup on its May 1969 cover with the title, "The final decline and total collapse of the American avant-garde."

The cover was designed by George Lois. Lois was a highly regarded American art director, designer, and author. He's in the Art Directors Club Hall of Fame, the American Advertising Federation Hall of Fame, and is also known for the 92 *Esquire* covers he designed, 38 of which are in the Museum of Modern Art's Modern Collection. MTV's *I want my MTV* campaign is his.

There's a quote from Warhol that he drank Campbell's Soup for lunch for twenty years. Did the tomatoes that Campbell's bred here fuel Warhol's creativity? I like to think so.

The iconic American comfort food combination, grilled cheese and tomato soup. Think about all the cans of Campbell's Tomato Soup that went into that combination. A lot of Jersey tomatoes were in those cans.

The Jersey peninsula though existed a long time before any food producers set up shop. New Jersey is the traditional territory of the Nanticoke and Lenni Lenape, called "Lenapehoking."

Like all the Indigenous people in the colonized world of Christopher Columbus, the Nanticoke and Lenni Lenape were driven from their land here and their land was stolen. Lenapehoking consisted of eastern Pennsylvania, southern New York, New Jersey, and Delaware. The northern part of Lenapehoking was inhabited by the Munsee (People of the Stony Country), while the central and southern areas were inhabited by the Unami (People Down-River) and the Unalachtigo (People Who Live Near the Ocean).

Considered to be one of the oldest of the Northeast Nations, the Lenni Lenape were peacemakers when disputes between nations broke out. The colonizers of the 1600s called the Lenape "Delaware Indians."

The Nanticoke lived in Southern Delaware and migrated from the Eastern Shore of Maryland. They were one of the first of the Northeast Nations to resist the colonizers. In the early 1700s, the colonizers forced the Nanticoke and Lenni Lenape onto reservations, which they claimed would protect them.

By 1758, the Brotherton Reservation, New Jersey's one and only reservation, was formed. It failed in its intended mission and was disbanded in 1802.

By this point, the nations were struggling to maintain community as their land was being stolen and Western diseases were taking their toll. The first treaty signed by the U.S. government after the Declaration of Independence was with the Delaware Indians. It promised statehood if the Delaware Indians would fight along with the government overseeing the war, but that promise was not upheld.

It was at this time, in the late 1790s, that the Nanticoke and Lenni Lenape were pushed north and west of their native land. Starting in 1860, the U.S. government's Indian removal policy following Andrew Jackson's Indian Removal Act of 1830 forced most of the remaining members of the two nations to Oklahoma, Wisconsin, and Ontario, Canada. This is a scare on the history of the United States.

While living on their land before the colonizers stole it, the Nanticoke and Lenni Lenape believed in a spirit that lived in the Jersey Pines called M'Sing. It was deer-like, with wings like leather. It tracks with the most common description of the Jersey Devil said to roam the Pines. The magical, enchanting, and bewitched Jersey Pines: the crown jewel of the Jersey peninsula.

The Pines hypnotize me while visiting. The folklore has a Jersey sense of humor and spirit running through it. Reading it while sitting in the Pines on a sunny fall afternoon is an energy charge like no other.

The Pines are home to an admirable individuality and self-sufficient generational history. This is also reflected in the stories of the Pineys, the locals in the Pines.

It is said that our devil, the Jersey Devil, is the thirteenth child of Mother Leeds. Mrs. Leeds had twelve kids, and when pregnant with the thirteenth, she cursed the kid by saying they would be the devil.

And as said, so it was.

It was a stormy 1735 night in the Pines. Mother Leeds was being attended during her delivery by friends when a normal-looking child was born. Then, in a flash of lightning, the child transformed into a deer-like creature with hooves and a forked tail.

In a frenzy filled with growls and screaming, that forked tail killed everyone, and the Devil disappeared up the chimney to go and live in the Pines and in the imaginations of campers sitting around fires.

The Internet, that storytelling oracle that repeats the same story, verbatim, without a single attribution, proclaims that in 1938 the Jersey Devil was designated the country's only state demon. Right on, another proud example of Jersey innovation.

Although I can't find a reference that supports that, what I could find is from Atlantic County, NJ's website:

> In 1939, the New Jersey Devil was reportedly named the Official State Demon.

Walter Edge, twice governor of the state, was quoted as saying:

> "When I was a boy. . . I was never threatened with the bogey man. . . we were threatened with the Jersey Devil, morning noon, and night."

I guess when you turn from a normal baby into a hooded creature with a forked tail, you get a spooky rep and the opportunity to befriend the spirit of a headless pirate. The infamous Captain Kidd is reputed to have buried treasure in Barnegat Bay, NJ, at the edge of the Pines. Evidently Kidd beheaded one of his men to guard this buried treasure. Witnesses claim the headless pirate and the Jersey Devil became friends and have been seen in the evenings walking along the Atlantic Ocean and in nearby marshlands.

Captain Kidd and other pirates like Blackbeard have their place in Jersey history. They sailed our waters. When I see the Mullica River, which runs through the Pines, I can imagine a vestige of pirate ships sailing the river.

The Pines are the largest remaining example of the Atlantic coastal pine barrens ecosystem, stretching across more than seven counties of New Jersey. Two other large, contiguous examples of this ecosystem remain in the northeastern United States: the Long Island Central Pine Barrens and the Massachusetts Coastal Pine Barrens.

White Cedar is found throughout the Pines. Tannins from them, along with a high concentration of iron, dye the water red in spots. Pitch Pines are the primary example of the evolution of the Pines. They can tolerate the droughts, fires, and acidic sandy soil carpet of the Pines.

They're a key food source for deer, rabbits, mice, birds, and insects. They have both male and female flowers and are wind-pollinated to create the cones that produce the seeds. Their complexity and beauty are composed of many fascinating evolutionary traits.

Oaks are the second most prominent tree in the Pines. The Jersey Pines are also home to three wetland flowers that have been thought to be virtually gone elsewhere but are locally abundant. They are the Pine Barrens Gentian, Bog Asphodel, and Swamp Pink.

Like blueberries and cranberries? Well, thank Elizabeth Coleman White and Elizabeth Lee, both Piney women and from farming families, for their roles in creating the commercial blueberry and cranberry markets. White worked with USDA botanist Frederick Coville on her family's cranberry farm to develop the first commercial blueberry bush. Lee was a farmer who made and sold the first New Jersey cranberry sauce, Bog Sweet, in 1917. She later worked with Marcus Urann of Massachusetts to form a cranberry co-op that became Ocean Spray.

The Pines are truly a special place. I went to college in the Pines. Stockton University's main campus is there. I lived in the Pines for a semester on campus, and in the couple times in my life that I worked in Atlantic City, which is on the coastal side of the Pines, I was still connected.

I've reconnected with the Pines lately, since a lot of great homesteading and food security work goes on there today. It's a continuation of the self-sufficient nature of Pineys.

That term carries a prejudice with it. Modern Pineys are reclaiming the Piney term, which had become shorthand for being ignorant, inbred wood dwellers.

True confession: growing up as a suburban transplant in a brand-new, privileged community that had a swimming pool, a nine-hole golf course, an apartment complex, its own elementary school, public basketball courts, and a tennis court, I bought into that awful trope.

I was wrong, and I apologize for it. I own it completely without qualifications.

Instead of taking the time to learn on my own about the generational stewards of the land, this incredible bio diverse masterpiece by Mother Nature, I bought into ignorance.

It wasn't until I got to Stockton and away from where I grew up that I learned the truth about how misguided and ignorant I was.

As much as I love Jersey, I have no illusions that we're above the worst traits of being human. We just do the worst bits as we do everything else, in our Jersey way. In 1913, Governor James Fielder called Pineys "*NJ degenerates.*"

He did this because of a book by Dr. Henry Goddard and Elizabeth Kite, *The Kallikak Family: A Study in the Heredity of Feeble-Mindedness.* Goddard was a eugenicist, and this book legitimized eugenics in the early 1900s. It was a study of the genealogy of a resident in Goddard's institution, the New Jersey Home for the Education and Care of Feebleminded Children in Vineland, NJ.

The gist was that he found a relative who sired a child with a barmaid after a one-night stand. Goddard claimed the descendants of this child were feebleminded, while descendants of a traditional valued line weren't. The Pineys, of course, were the feebleminded descendants.

This book has since been debunked, but the generational implications exist to this day, both here in Jersey and in society in general. Selective breeding is okay for plants, but for humans it's as evil as it could be.

The isolation of the Pines goes back to the colonizer days, when they called the Pines, the Pine Barrens because they couldn't grow their crops there. The Pines are anything but barren. Early Pines industries include bog iron, an impure form of iron that's mined from bogs, with about 35 furnaces throughout the Pines. Iron from the Pines supplied American military weapons and tools in the Revolutionary War, the War of 1812, and the Second Barbary War. Shipbuilding, sawmills, cotton mills, and cranberry bogs were also key industries.

The diverse ecosystem that exists there is a wonder. It's fortunate that this area has been preserved from development thanks in part to the 3,000-square-mile Kirkwood-Cohansey aquifer. After reading John McPhee's fourth book, *The Pine Barrens,* in 1977, then-Governor Brendan Byrne preserved the Pines with an executive order.

The Pines' ecological diversity is matched by Jersey's population. We're one of the most diverse states in the country. In a study conducted by WalletHub, Jersey City was ranked the most diverse city in the country in 2017, 2018, 2020, and 2021. This study compared ethnic diversity, ethno-racial diversity, language, and birthplace in more than 500 cities. The same study listed Jersey as the fourth most diverse state in the country.

When Bruce Springsteen romanticized summer nights along the Jersey Shore in his version of "Jersey Girl," he introduced the public to a working-class single mother. Springsteen adds another layer of his Jersey working-class ethos that's demonstrated universal appeal. With Springsteen, there's hope in the darkness and the challenges of the working class.

When he sings about how the shore makes everything all right, the hope here is that a night of self-love filled with the warmth of human interaction and the warm, sultry Jersey Shore will make everything all right. The warm and breezy summertime Jersey shore is a reset. We know that here and celebrate that as part of our culture, our place here at the eastern edge of our peninsula.

This hope is centered in what Springsteen and Tom Waits, the original author, both know: how very special Jersey Girls are. Not many songs in the Great American pop/rock playbook honor women from specific states, named in the title.

We're a spirited bunch here. Yes, there's the memes of Jersey Girls with big hair and acid-washed jeans, but Jersey Girls embody the bold and passionate nature of our Jersey culture. They're unapologetic, independent, and loyal. They are who you want on your side.

Just don't mess with them or your table could get flipped. When Teresa flip the table in *The Real Housewives of New Jersey,* I knew that was a true Jersey Girl move.

Another true moment was from Dionne Warwick in her *The Dionne Warwick: Don't Make Me Over* documentary. When Marlene Dietrich was tossing her dresses off a wardrobe rack in a hallway, she said:

> *"I thought the woman lost her mind. The New Jersey stared coming out from me."*

Jersey Girls are authentic. They know who they are. While I can't directly relate to the romance being professed by Waits and Springsteen, since I'm gay, I can attest to a platonic love because authenticity is the key to my heart and it exists in the heart of Jersey Girls.

That authentic Jersey heart and soul is captured in *Doonesbury's* Lacey Davenport. While Garry Trudeau denies the connection, it's thought that he based her on Millicent Fenwick, a trailblazing NJ politician. I read *Doonesbury* religiously, when Fenwick was in Congress while I was growing up; the similarities are very close. I can see the Jersey Girl spirit in both that gives life to that connection.

The Sopranos have been called one of the most influential TV shows in history. Some credit it with reviving a lagging interest in a television series. And it was set in Jersey. It's an interesting choice to show a mobster outside of a city, which is usually associated with mafia power. David Chase, the creator of the series, drew heavily on his life growing up in Jersey. He tried to apply that and his family dynamics to mobsters.

Chase grew up in Clifton and North Caldwell, NJ. James Gandolfini, who played Tony Soprano, grew up in Jersey—Park Ridge, to be exact. So he had the Jersey in him when he played Chase's Jersey mob boss.

The Sicilian Mafia migrated with Sicilians to America, and eventually became the American Mafia, or Mob. Being second-generation Italian, my family circles were concentrated in South Philly. My family wasn't connected, but when you travel in circles of first- and second-generation Italians migrants, it's hard to avoid it.

Where I grew up in South Jersey, it was an enclave of former South Philly residents. So, like the Sicilian Mafia moving from Sicily to America with migration, the mafia from South Philly made its way to the enclave where I grew up.

And the Jersey mob eventually took on the edge that comes with living in Jersey, which Chase reflected in his character played by Gandolfini. Add in the beloved Pine Barrens episode, and the Jersey influence on this best-of-its-kind American pop culture series and unforgettable characters is complete.

While in Vancouver, right after *The Sopranos* ended, I was told by some guys at a gay bar that my Tony Soprano accent was sexy. Imagine that, the exotic foreigner with the accent, and it's Tony Soprano, the made man from Jersey.

No look at Jersey's place in popular culture would be complete without the backdrop for Kevin Smith's work, the *View Askewniverse*. It is a fictional universe based on two towns in Monmouth County, NJ, where Jay and Silent Bob live on forever. While some who visit or move here might feel that Jersey is a fictional universe, I can assure you it's not. You can't make Jersey up. Just take bits of inspiration and amp them up to showcase the Jersey way of the idiosyncrasies of the human experience.

Smith owns Jay and Silent Bob's Secret Stash in Red Bank, NJ. It's a comic book store with a collection of Smith's merchandise. Jersey roots run deep.

Nothing captures NJ more than *Weird NJ*. A local travel guide published since 1989, it will take you off tourist maps to places uniquely Jersey.

Zach Braff's *Garden State* is a well-received film about an actor returning home to New Jersey when his mother dies. New Jersey provides the backdrop for this story which is partly autobiographical.

Atlantic City (AC), The Queen of Resorts, has a dramatic history closer to a saint's than a cloistered monarch. AC is where I danced my way into my life, as a gay man at the Saratoga on New York Ave., and it retains a special place in my heart.

Once air travel to Florida and the Caribbean became more accessible in the '50s, AC lost its prominence as a premier resort destination. This Jersey Shore gem faced an economic abyss. Casino gambling was pitched as a way to revitalize AC. Voters approved the plan, and New Jersey became the first state outside of Nevada to offer casino gambling on November 2, 1976.

The results have been mixed at best. I never worked for a casino. I worked in and around the casino industry in AC. I also attended Stockton, just outside AC, for four years shortly after gambling was approved.

I was ahead of the curve on taking a gap year after college before starting my career. My first stint was as a breakfast waiter, where shortly after, I was promoted to bartender in a non-casino hotel located within walking distance of three casinos at that time: Bally's, the Sands, and the Claridge. The three casinos were clustered around Brighton Park, and all were within a block and a half of each other. Bally's is the only casino still there. The Sands was demolished for a proposed resort that never happened, and is now a vacant lot. The Claridge operates as a hotel only, and was recently approved for an adult cannabis dispensary and adult use lounge in its former casino.

The hotel I worked in catered to bus tours. Casino bus tours became a large part of early gambling in Atlantic City. These folks would get $20 in quarters and a meal credit as a draw to the casinos. The hotel I worked in would have overnight or two-night package deals for drinks and food at the restaurant, along with some small premiums from the three casinos that were so close together.

I met a lot of people from West Virginia, Western Pennsylvania, Massachusetts, and Maryland. It was my first job in a restaurant. I went from being a breakfast waiter, to having the best money shifts behind the bar in a matter of a couple months! That gap year showed me more career title advancement in a short time than any corporate job after.

The bar was also the de facto lounge for dealers from the Sands and Claridge casinos. Occasionally some table dealers from Bally's would come over, too. At that time casinos closed at 4 am Sunday-Thursday and 6 am Friday and Saturday. What I learned from this experience was that the folks taking the buses just wanted to gamble. They didn't care about engaging with Atlantic City in any other way, including the renowned Boardwalk.

The regular casino employees that came to the bar said the same thing on the casino side of it. The goal was to keep them inside the building, so the promised redevelopment never happened.

Money was kept in a very tight circle that didn't expand outside of the casino properties. The excise tax that was part of the package the voters approved did little to help the community and neighborhoods in AC.

This set-up was perfect for the business model of Donald J. Trump, who made a big splash investing in AC, only to go bankrupt, leaving behind three failed casinos and many invoices from local contractors unpaid.

When Trump arrived I was working for a small local tabloid publishing company. We did three monthly publications: one a trade journal targeted to casino employees, one was targeted to gamblers, and one focused on the local lifestyle. I was privy to the content that was relevant to the casino industry on the business side and on the consumer side.

Again, the focus was on keeping folks inside the casinos, and keeping the money away from the community.

I did get to a Miss America pageant, and it was a lot of fun. It was just one big party. There were many bars along the perimeter of the old Convention Hall, now Boardwalk Hall.

When the commercial breaks happened there was lots of laughing and a party buzz, which all quieted down once the network was back on. Miss America helped define AC when it was launched in 1921. The Miss America contestant parade on the Boardwalk became known as the Show Us Your Shoes Parade in the '70s. It was either from a drag queen named Tinsel Garland crashing the parade dressed as Dorothy from *The Wizard of Oz* (the crowd chanted "Show us your shoes" and she would show her ruby slippers), or from guests on a balcony looking down into the convertibles and seeing no footwear on the contestants. Either way, it became a beloved part of the parade while it was in AC.

Steel Pier had the diving horse, where a horse would dive directly into the ocean, and the diving bell, where folks could pay an admission to descend into the ocean in a steel diving bell. It had small windows and microphones connected to speakers on the pier so you could send messages to your friends on the pier from the depths of the Atlantic Ocean. The only underwater view you had was of the pilings the pier was built on. Lots of good music played on Steel Pier, including Diana Ross and the Supremes for week-long stints in the mid '60s.

The television show *Boardwalk Empire* and the board game Monopoly, which was originally based on the streets of AC, weave AC into the fabric of American culture. Honeymoons, a glassblowing studio dating back to 1901 on the Boardwalk, boxing, concerts, White House subs, Formica Bakery sandwich rolls, the beach, saltwater taffy, and boardwalk pizza are all part of AC. While the diving horse is gone, diving seagulls that want to steal your pizza still exist.

Beware.

So now that you have a sense of who I am and my view of Jersey terroir, culture, and history from my life experience here, it's time to talk tomatoes, the intersection of it all in this magnificent story.

A lot of the local conversation around Jersey tomatoes laments their lost flavor. When you ask folks what variety of tomato they are complaining about, they can't tell you. Jersey tomatoes do have, a sweetness, and a distinct acid note. They have a meaty texture, and there is a nice level of juice that all combine to create the flavor experience.

These tomatoes were bred for canning; when heat is added, complexity of flavor amps up. With each variety having its own flavor, mixing the varieties, you can create subtle flavor differences.

A lot of farmers will tell you it's not the variety that matters, it's the soil and weather. That's a very good point, since terroir has a direct impact on flavor, plant disease, and production.

The climate has changed, and over the years the soil has been impacted by the use of chemicals, erosion, and a general approach to agriculture that doesn't treat soil holistically. Organic and regenerative growing do, but New Jersey isn't a leader in organic growing. We're getting better.

Then there's the development of hybrids that have made for stronger disease resistance and production, which are very important considerations for farmers who need to make money.

Hence, the laments about flavor. One of the main reasons I think this conversation happens is that a lot of farmers were growing for The Campbell Soup Company, or Campbell's, when their production was here. Campbell's bred for taste along with high production and disease resistance. They provided plants to the local farmers, and employees. They only accepted the best of the harvest. The rest were likely sold at farm stands, farmers' markets, or food markets that sold local produce (unlike most supermarkets today).

Once Campbell's moved their growing to California and their soup production to other processing facilities outside of New Jersey, so did their plant distribution and the tasty tomatoes they bred. There are other Jersey-bred flavorful varieties, but they all fell by the wayside once the seed companies here left.

At one time tomatoes were thought to be poisonous. There is local lore about how that changed in Jersey. On a bright sunny day on the Salem City Courthouse steps, either on June 28, 1820, or September 26, 1820, Colonel Robert Gibbon Johnson stood and, in a truly brave and death-defying act, ate tomatoes to dispel the myth that they were poisonous. There's no documentation that this actually happened, hence the conflicting dates. And by that point, tomatoes were grown and consumed across America.

From the Salem County Historical Society:

> Col. Johnson announced that he would eat a tomato, also called the wolf peach, Jerusalem apple or love apple, on the steps of the county courthouse at noon. ... That morning, in 1820, about 2000 people were jammed into the town square. ... The spectators began to hoot and jeer. Then, 15 minutes later, Col. Johnson emerged from his mansion and headed up Market Street towards the Courthouse. The crowd cheered. The fireman's band struck up a lively tune.
>
> He was a very impressive-looking man as he walked along the street. He was dressed in his usual black suit with white ruffles, black shoes and gloves, tricorn hat, and cane.
>
> At the Court House steps he spoke to the crowd about the history of the tomato. ... He picked a choice one from a basket on the steps and held it up so that it glistened in the sun. ... "To help dispel the tall tales, the fantastic fables that you have been hearing ... And to prove to you that it is not poisonous I am going to eat one right now"... There was not a sound as the Col. dramatically brought the tomato to his lips and took a bite. A woman in the crowd screamed and fainted but no one paid her any attention; they were all watching Col. Johnson as he took one bite after another. ... He raised both his arms, and again bit into one and then the other. The crowd cheered and the firemen's band blared a song. ... "He's done it", they shouted. "He's still alive."
>
> — "The Story of Robert Gibbon Johnson and the Tomato," Salem County Historical Society

Johnson was also a slave holder. The tomatoes that he claimed to have defied death with were germinated, grown, and picked by enslaved people. In 1826, there was an incident with Johnson's second wife and the enslaved Hetty Reckless, who fled to a safe house in Philadelphia. What I've read said Johnson's second wife abused Reckless. She fled with her daughter and became an abolitionist, and did not return to Salem until after Johnson's death in 1850.

Reckless story is documented. Johnson's death-defying act isn't. A complete history must be told, especially when one part of it is documented, and one is lore. The main character in the lore, a wealthy white man, walked around in a black suit with white ruffles, black shoes, gloves, tricorn hat, and cane, is not documented. And the other character, was enslaved, abused so much that she escaped, and became a leader in a movement to free other enslaved people, is documented.

But you know. white ruffles, and a tricorn — very fine.

Sonya Harris is a friend and garden educator. Sonya might be more passionate about Jersey tomatoes than I am, is tried-and-true Jersey, and of the most beautiful humans I've met. A true Jersey Girl.

We were talking about this charade, and the role of slavery on New Jersey farms, and specifically the area where we live came up. While technically in the North during the Civil War, the Mason-Dixon line passes through our state. The history of slavery in New Jersey is not talked about much here, despite the fact that we were the last slave state of the North and the legislature in March 1865 initially refused to ratify the Thirteenth Amendment.

That changed on January 23, 1866, when a constitutional amendment was signed by Governor Marcus L. Ward. It was his first act as governor, months after Juneteenth. Jersey did play a significant role in the Underground Railroad, and that's talked about, but not the dark side of our slave history as demonstrated by the Colonel Johnson lore. The complete story has to be told about it. Slavery is a dark part of history, and it impacts all our social structures. It's certainly not connected in the public eye to Jersey tomatoes.

Seed work consists of correcting the record about seeds, and reframing Robert Gibbon Johnson's story from Reckless's perspective is a prime example of that work. The toxic masculinity in this local tomato lore—if it even happened—was, in reality, a stunt that further exploited Johnson's enslaved people, since he wasn't the one doing the hard work of germinating, planting, or harvesting the tomatoes he so heroically demonstrated were safe to eat.

Enslaved people were.

Fourteen Jersey Tomato Histories

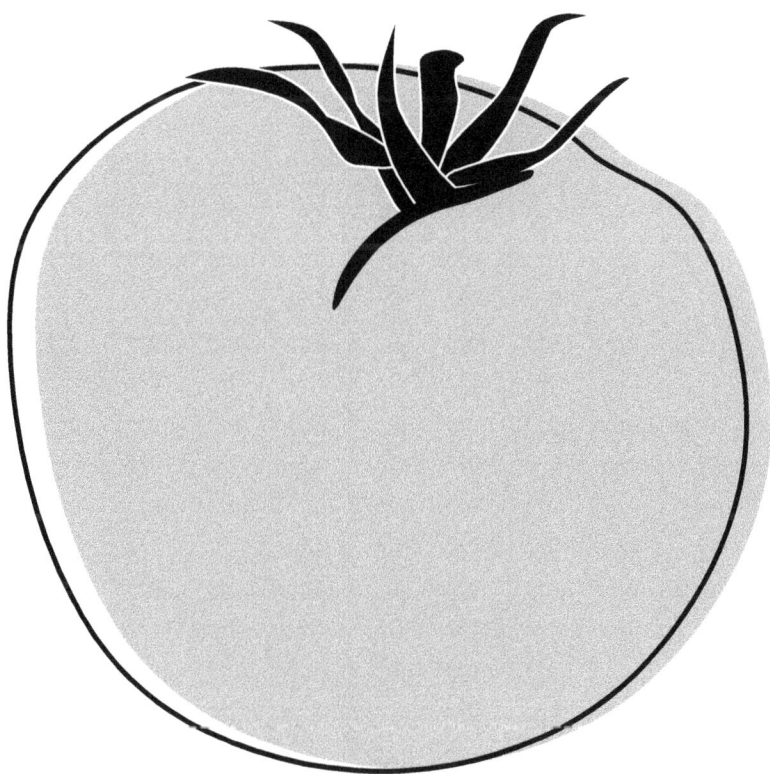

Sparks Earliana

Originator:	George Sparks
County:	Salem
Introduced:	1900 Johnson & Stokes, Mooerstown, NJ

So, if the iconic Jersey tomato is a product of its terroir, then it would make sense that a genetic mutation in a field of Stone tomato plants in the fertile soil of Salem County, NJ, would be the original Jersey tomato variety.

It's so Jersey. Since we're Jersey and have our way of doing things here, the as my friend Leah Klem once said, *"Sparks Earliana tomato said get out my way, this is how it's done."*

When George Sparks of Perkintown, Salem County planted his fields in late 1897, little did he know how his field of Stone tomatoes would change his life, the direction of the tomato business, and Jersey tomato history.

From the article *"Sparks Earliana,"* Salem County Historical Society:

> In 1899, the editor of The Penn's Grove Record made the following observation: "Two years ago Geo. Sparks found one hill of a new variety among his early tomatoes. [Sparks had planted a field of tomatoes of a variety named "Stone."] He saved the seed and last year he planted them. They proved to be a fine tomato, very productive and an extra early one. This year his whole crop is of that variety. Between the 14th of June and the 22d of July he had received returns amounting to over $600.

A seed company takes note:

> "A member of the firm of Johnson & Stokes, of Philadelphia, came down to Mr. Spark's farm near Perkintown and procured a plant, photographs of which will doubtless appear in their 1900 catalogs, as the firm will put the seed on the market next year. ..."

Johnson & Stokes Seed Company was that seed company based in Moorestown, NJ. The name of the tomato they brought to market is a mash-up of "Earli," because it was the earliest market tomato of that time, and "Ana," Herbert W. Johnson's—a partner in Johnson & Stokes—granddaughter. The seed was released in 1900.

Sparks Earliana made such a splash after it was introduced that in 1901, a year after its introduction, it caught the attention of W. Atlee Burpee of the Burpee Seed Company. He included it in his *Novelties of 33 Tried and True New Vegetables for 1901*:

> *This is the earliest smooth bright-red Tomato of good size now in cultivation. It has been developed in Southern New Jersey where the greatest competition exists among growers to be the first in the market. MR. GEORGE C. SPARKS, of Salem Co., has been most successful in developing his "Earliana,"…In our Trial Grounds a specimen plant had forty-five ripened fruits at time of the first picking and thirty-five green ones of good size and yet the whole plant could be completely hidden by an ordinary bushel basket.*

That's a lot of tomatoes on a plant.

Two years later, from a two-page victory lap spread, these headings appear from in Johnson & Stokes' *Garden and Farm Manual, 1903*:

> *The Greatest Novelty in Tomatoes Ever Offered*

> *THE EARLIEST LARGE TOMATO IN THE WORLD*

In a lot of ways, seedsmen come off as carnival barkers. Johnson & Stokes, in this case, had every right to boast. They were the keepers of the original seeds of the strain of the Sparks Earliana.

The copy closes out with this line:

> *Seed crop very short. To be sure of genuine seed, grown direct from stock seed selected by Mr. Sparks, the originator, order direct from the introducers.*

Lineage is very important in seeds, and it's good to see that stated on this page.

The illustration is of a man admiring a large tomato tracks. I would be smiling too if I was holding the most profitable tomato ever grown.

This illustration first appeared on the cover of Johnson & Stokes' 1902 catalog in color, but on this page, it was a simplified version of just the man without the background that was on the cover.

The back cover of the Johnson & Stokes *Garden and Farm Manual, 1903* featured another illustration about the Sparks Earliana. It repeats the headline, *"Sparks' Earliana Has Proven the Most Profitable Tomato Ever Grown,"* and on the bottom of the page, it has this:

> The earliest and most prolific large tomato in the world. First
> introduced by Johnson and Stokes Seedsman. 217 and 219 Market
> Street, Philadelphia, PA.

While their business office was in Philly, they had a large experimental farm in Moorestown, Burlington County, New Jersey, which is about 35 miles from Perkintown, the site of George Sparks' farm. Local seed is the best.

Sparks Earliana became a mainstay of the tomato canning industry in New Jersey. I found a reference in a soil survey in 1918 that indicated the Sparks Earliana was being grown. I've seen a Penn State version referenced in a Pennsylvania Ten-Ton Tomato Club document from 1943. The Ten-Ton Tomato Club, started by the New Jersey Canners Association, was a tomato growers' group with an award program.

If a farmer pulled ten tons of quality tomatoes off their land, they were recognized with an award. There was a Five-Ton Club as well for the five-tonners. Pennsylvania had their own version of the club, since the eastern counties that surrounded Philadelphia grew a lot of tomatoes.

From that hill of 4,000 plants in Perkintown, magic happened, with a genetic mutation in Jersey soil. That magic set the standard for Jersey tomatoes.

The seed company that introduced it was 35 Jersey miles away.

Pride and passion in tomatoes are true Jersey traits. And what better representation of that pride and passion than the fact that, in the early 1900s, when the canning industry was taking off here, inside those cans was a true Jersey original tomato.

Photograph of a single plant of SPARKS' EARLIANA TOMATO, showing its wonderful productiveness.

The Greatest Novelty in Tomatoes Ever Offered

SPARKS' EARLIANA

THE EARLIEST LARGE TOMATO IN THE WORLD

Very early tomatoes have of late years been such a profitable crop that almost every grower in Southern New Jersey boasts, with more or less justice, that he has the earliest tomato. We have for years watched with a good deal of interest the first shipments to reach Philadelphia markets, and were not a little surprised to find, three years ago, a new variety from Mr. Geo. C. Sparks in market fully two weeks ahead of all others. We immediately arranged with Mr. Sparks to save us some selected seed, paying him what was probably the highest price ever paid to a grower for the control of a new tomato, and which we distributed in a small way among our customers. Nothing we have ever introduced has brought us so many strong testimonials, a few samples of which we publish on the opposite page. This tomato is not only remarkable for its earliness, but for its very large, uniform size, handsome shape, beautiful red color and wonderful productiveness, standing unequaled and alone in this respect. Its solidity, fine size and quality are unsurpassed.

A single cluster of SPARKS' EARLIANA taken from our field.

Seed crop very short. To be sure of genuine seed, grown direct from stock seed selected by Mr. Sparks, the originator, order direct from the introducers. Pkt., 20c.; oz., $1.00; 4 ozs. for $3.50.

Sparks' Earliana Has Proven the Most Profitable Tomato Ever Grown

Realized $200 on Less than One-half Acre

R. E. LEE BRADFORD, Fort Worth, Texas, August 2, 1902, writes: "I sold $200 worth of tomatoes from less than one-half acre of your Sparks' Earliana. It beats any tomato ever grown for this market in earliness and productiveness. I will plant three acres of them next season."

Over $300 from 1500 Plants

C. M. EMERORY, Knoxville, Tenn., Oct. 17, 1902, writes: "The ½ ounce Sparks' Earliana Tomato Seed purchased from you gave me 1500 plants from which I made over $300. I have been growing tomatoes for 18 years and have never seen its equal. Fine large tomatoes more than three weeks ahead of all others on the market."

$750 on Less than Two Acres

MR. JOSEPH STILES, a well-known trucker at Pennsgrove, N. J., says: "I realized $750 on less than two acres of your Sparks' Earliana Tomatoes."

$6,000 is Cleared by a Norfolk Trucker

MR. NIMROD C. CROMWELL, one of the oldest and best known truckers at Norfolk, Va., came up from Norfolk to our store in Philadelphia to purchase some seed of the Sparks' Earliana Tomato, and while there stated to us that "his two sons, who are now running his farm for him, had cleared over $6,000 on Sparks' Earliana Tomatoes the past season."

$725 Clear on 4,000 Hills

MR. GEO. C. SPARKS, the originator, says that "in the year 1899, when I first marketed the Earliana Tomatoes, I realized from my first pickings $725, clear of freight and commission, on about 4,000 hills."

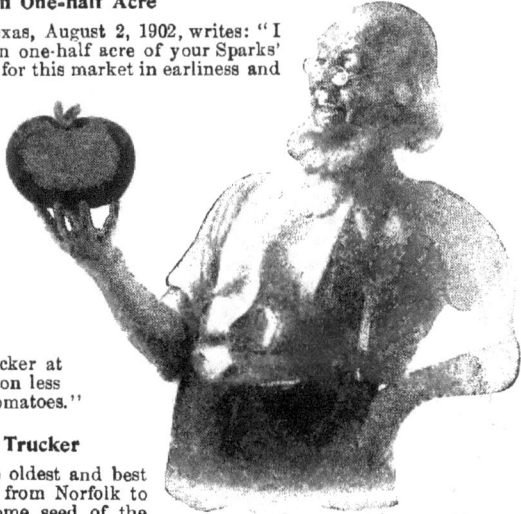

What a Few of the Largest Market Growers Say of Sparks' Earliana

BEAT ALL RECORDS BY SIX WEEKS

JOSEPH A. WEIS, North Adams, Mass., July 8, 1902, writes: "Your Sparks' Earliana beat all records here by six weeks. Many gardeners have not yet brought a ripe tomato to market. People scarcely believe I raised mine so early; they think I had them shipped to me."

BEST IN THIRTY YEARS' EXPERIENCE

WM. H. KIRBY, Chestertown, Md., June 21, 1902, writes: "Your Sparks' Earliana is the finest tomato I ever saw. It is the earliest and most productive. In the past thirty years I have tried many varieties of tomatoes, but this one beats them all; yesterday one of my men counted 176 tomatoes on one plant."

NO OTHER WILL BE GROWN

WARREN WEBSTER, Port Monmouth, N. J., September 9, 1902, writes: "I was the first to introduce in this neighborhood, in 1901, your Sparks' Earliana Tomato; next Summer there will be no early tomato but Earliana here, for its merits are now known to all. They are large, good shippers, good sellers, enormous croppers, and earlier than any tomato we have ever seen."

ALL THAT COULD BE WISHED FOR

GEO. H. GILL, Little Neck, L. I., November 1, 1902, writes: "I had 7,500 hills of your Sparks' Earliana and they were by far the earliest and best I ever raised. In size, earliness, and productiveness they are all anyone could wish."

MOST VALUABLE NOVELTY IN A CENTURY

JOHN W. MILLETT, Bismarck, N. Dakota, writes: "I congratulate you on having introduced the most valuable tomato novelty of the past century, which Sparks' Earliana surely is. It is earlier than all other varieties ever brought forward; it produces three times the amount of fruit of any variety I ever grew. This superb tomato not only equals but far surpasses all you claim for it."

EARLIEST AND BEST OF 100 VARIETIES

ALFRED PALMER, Salem, N. C., July 5, 1902, writes: "After forty years' experience in tomato raising I must say that of over 100 varieties I have tried, I found your Sparks' Earliana to be the very earliest and best in every way. Its quality and merits are perfection."

WITHSTOOD WET WHEN OTHERS ROTTED

C. W. MACMILLAN, Savannah, Ga., January 28, 1902, writes: "I found your Sparks' Earliana everything you claim for them; not only were they heavy bearers and at least ten days earlier than any other varieties planted at same time, but they withstood an exceedingly wet season without damage, whilst other kinds rotted badly."

HEAVY CROP WHEN OTHERS FAILED

GEO. WOLFRUM, JR., Kresson, N. J., October 2, 1902, writes: "I have grown your Sparks' Earliana Tomato since its introduction, and find it the very earliest and best. Last year I had a heavy crop of them while my neighbors had hardly any from other varieties."

RIPE TOMATOES IN JUNE IN MARYLAND

MARY MAGRUDER, Olney, Md., August 9, 1902, writes: "People around here are delighted to have tomatoes in June of so fine a quality as Sparks' Earliana; the earliest this year were gathered June 21."

EARLIEST AND HEAVIEST CROPPER

HALLOCK JOHNSON, Villa Ridge, Ill., October 15, 1902, writes: "The Earliana is of good size and of the very best quality. It is the earliest of all early tomatoes, and the heaviest cropper I have ever seen grown."

AVERAGED 150 PER PLANT

J. T. JOLLEY, Boles, Ill, September 19, 1902, writes: "Your Sparks' Earliana Tomatoes got away beyond all you claim for them; they bore for me 150 to the plant on an average. A perfect wonder in earliness."

Sparks' Earliana is as near seedless as any tomato we ever saw, containing less than one-fourth the seed of other varieties, and for this reason the genuine seed can never be sold at a low price. We control the entire supply of seed grown direct from the originator's selected stock seed. Owing to the very short crop, we are unable to offer the seed in larger quantities than ¼ lb. in 1903. Prices, genuine seed: Per pkt., 20c.; oz., $1.00; ¼ lb. for $3.50.

known to them the contents thereof, they did then
acknowledge that they signed, sealed and delivered
the same as their voluntary act and deed, for the uses
and purposes therein expressed. H. K. Bughie. Com
Recorded May 17" 1907   Frank B. Ridgway
At 1.30 o'clock P. M.   Clerk b

Wilson J. Jones Sheriff.   This Indenture, made the
          to   twenty seventh day of December
W. Atlee Burpee.   in the year of our Lord, one
                                thousand nine hundred and
four Between Wilson J. Jones, Esquire, Sheriff of the
County of Gloucester, in the State of New Jersey, of
the one part, and W. Atlee Burpee of the City and
County of Philadelphia and State of Pennsylvania
of the other part. Whereas a certain writ of the State
of New Jersey, commonly called a fieri facias, issued
out of the Gloucester Circuit Court and to the said
Wilson J. Jones, Esquire, Sheriff of the said County
then being, was delivered, in which said writ it is
recited as follows: The State of New Jersey, Gloucester,
to wit: to our Sheriff of our County of Gloucester
Greeting: We command you as before we have that

| SEAL. |

of the Goods and Chattels of George S. Turner
defendant in your County, you cause to be
made the sum of Two thousand three hundred Dollars of debt, which Swedesboro National Bank
plaintiff lately before our Circuit Court holden in Woodbury in and for our said County of Gloucester recovered
against him and also Five Dollars, fifty cents, which
in our said Court before our said Judge were adjudged
to the said plaintiff for its damages which it had
sustained, as well by occasion of the detention of
that debt as for its costs and charges by it about
its suit in that behalf expended whereof there is
a recovery. And if sufficient goods and chattels
of the said defendant cannot be found in your
County, whereof the debt and damages aforesaid
may be made, then you are hereby further command-
ed, that you cause the whole or the residue, as the
case may require of the said debt and damages,

Burpee Sunnybrook Earliana

Originator:	W. Atlee Burpee
County:	Gloucester
Introduced:	1907 Burpee Seed Company, Horsham, PA

Oscar Wilde said, "Imitation is the sincerest form of flattery." W. Atlee Burpee wrote a lot of flattery in his seed catalogs about Sparks Earliana starting in 1901.

He also started growing it. At the Smithsonian Gardens, which have a large collection of Burpee's personal and business papers, I read all of his seed trial journal entries for the Sparks Earliana from 1901, when he began this grow-out, to 1906, the year before he released his own version. It generally takes five years to select and standardize the characteristics of a new variety before releasing it to the public. Noted in the 1901 Burpee seed trial journal for the Sparks Earliana entry is the word "Isolate." It's written in red and underlined. This is important to maintain genetic purity.

Burpee released his version in 1907, which makes sense because of the five years it generally takes to stabilize the genes so the variety is open-pollinated. Open pollinated is a term that means the seeds are true to form. You can plant and save the seeds and there won't be any deviations.

Sunnybrook was the name of Burpee's experimental farm in Woolwich Township, NJ, in Gloucester County, where he grew and selected for his Burpee Sunnybrook Earliana. Burpee continued his tradition of putting the name of the experimental farm in a variety's name. Fordhook is another Burpee farm, and that name is used in some Burpee varieties. Woolwich Township, is right next to Swedesboro. Both share a proud history of seeds, farming, and canning.

My community garden plot while I was discovering this history was in Woolwich Township, NJ. In 2021, I read those seed trial journals when I was reviving the Sunnybrook Earliana. I knew that Burpee's Sunnybrook farm was in Woolwich, I just didn't know where until I tracked down the deed from the purchase of the land that would become Sunnybrook Farm.

My plot was less than two miles from where Sunnybrook Farm was. When I started this process of living and recreating this history, I had no idea of the wonder and amazement that I would experience. To read the experiences of the seed farmers growing and selecting for a variety while reviving their work in very close proximity to where they worked blew my mind. I don't think I'll ever realize the full impact of doing that.

I have also seen the 1906 Sunnybrook crop list. It lists one acre of Sparks' Earliana. An acre would produce enough seed too for the release of the Burpee Sunnybrook Earliana in 1907. This list is also from the Smithsonian Gardens. I'm grateful for their records, and for Joyce Connolly, the archivist I worked with, who helped me piece this history together.

I came across the Burpee Sunnybrook Earliana while searching for another Burpee variety. During one of my previous visits reviewing seed trial journals at the Smithsonian Gardens, I came across a melon called Spicy. The name caught my attention, and the entries written about it doubled my interest. One year, they devoted five acres of land to producing seeds for this variety. That's a lot of seeds.

I wasn't able to locate seeds for this melon online, so I went to the USDA Germplasm Bank and did a search for Spicy melon that didn't return any melon results. Maybe the word "Spicy" was a code name for the melon, or a melon they were growing for someone else.

Next, I did a general search for Burpee, and that's when the Burpee Sunnybrook Earliana said "Hello" in my search results.

My jaw dropped, since I knew Burpee had an experimental farm in the vicinity of Swedesboro/Woolwich called Sunnybrook Farm. This was before I found the deed. I also knew the history of the Sparks Earliana.

Perkintown is where Sparks Earliana introduced herself to George Sparks. It's about fifteen miles from Woolwich Township, the site of Sunnybrook Farm. We talk about preserving backyard biodiversity in seed saving, which means saving seeds from your own backyard since it's a microclimate, and seeds will adapt to that growing area. That's optimal for seed saving.

Tomato breeding is a big part of the local story of seed and agriculture history. It's part of our culture. So, while fifteen miles isn't a backyard these days, in the heyday of seed breeding and farming, before suburbs, it was.

I have a seed-saving project called Library Seed Bank where I brought the concept of seed libraries to South Jersey. I launched it in 2014. Over time, the mission has evolved to connecting the public to their local seed and agriculture history through seed libraries. This is accomplished by working with the communities where these seed libraries are located to build up a local supply of backyard, bio diverse seeds as close as possible to where they were originally bred, trialed, and introduced.

It's a homage to all the farmers, growers, and seedsmen who built seed sovereignty at a time when that action wasn't needed. It's through a seed and food sovereignty lens that I look at the local food history, here since it was such a solid representation of it. Seed and food sovereignty is your right to secure your own culturally appropriate seeds and food without interference. It's individual power over seeds and food, not consolidated power by governments or corporations. It also keeps the money local and creates a circular economy that supports stronger communities.

It was a way of life back then, but today social action is needed. The vast difference is where the power lies. The agricultural and food system is an industrial complex with power concentrated at the top.

So, with all this in mind, and knowing that my community garden plot at the time was in Woolwich Township, I personally had to grow this tomato. I would be remiss in my commitment to the activism that I live and breathe every day if I didn't. Also, the community would be missing out on the opportunity to revive a local tomato, and ensure it stays local now through seed saving and seed libraries.

The Gloucester County Library system has a robust seed library thanks to Jim O'Connor, including a branch in Woolwich Township, just down the road from where Sunnybrook was and where I revived this tomato. They also have a branch in Swedesboro, a one-time seed and canning hub.

For the research part, after I saw the results from the USDA search, my immediate response was to search for a Burpee catalog. I took a guess about the year, 1915, and downloaded a PDF of the *Burpee's Annual for 1915* catalog. On page 91, I found the Burpee Sunnybrook Earliana with the following closing copy in the description:

> *This Special Stock is grown exclusively on our Sunnybrook Farm, in Gloucester County, New Jersey and is only sold under our registered trademark.*

Bingo! This Burpee catalog cited Gloucester County for a tomato that he was proudly growing in our soil. So it's not just Jersey boasting, it's Burpee too, whose headquarters are in Pennsylvania.

I knew I would be able to bring this tomato back to Gloucester County where Burpee selected it, and that the seeds would be distributed through branches of the Gloucester County Library System, including the branches in Woolwich Township and Swedesboro, both down the road from Sunnybrook Farm.

Thanks to the power of seeds, I got to touch history with my hands and recreate this tomato in the soil very close to where it originated. That history is in a seed that grows what characteristics Burpee selected for. That doesn't change no matter how long the seed sits in a seed collection. It's not just this seed or variety. It's every seed.

I'm not really a fan of Burpee's registered trademark, by the way. I'm a firm believer in open-source seeds, where there is no legal claim to ownership of a seed or any of its traits. Mother Nature holds all trademarks as far as I'm concerned, but I don't believe she cares about them. Biodiversity is her gift to life on earth. One of the reasons I create seed libraries for the public, is that they offer free and open public access to seeds like Mother Nature intended. It's good to know that, with community partners, I've been able to create this permanent social change in South Jersey.

While reviving the Burpee Sunnybrook Earliana in my garden plot, I got to read from the journals about what was happening in Burpee's seed trials less than two miles from where I was growing it. This is part of the earlier Burpee quote:

> In our Trial Grounds a specimen plant had forty-five ripened fruits at time of the first picking and thirty-five green ones of good size and yet the whole plant could be completely hidden by an ordinary bushel basket.

I grew 25 plants. A lot of the plants were this prolific. The Sparks Earliana is known for how prolific it is, and this description from Burpee is accurate. My second tomato set got sterilized by the heat. Tomato pollen goes sterile when the temps are over 90 degrees Fahrenheit during the day and above 70 at night for longer than three or four days in a row. The nighttime temperatures are most critical though.

In July 2021, in a week I went from seeing my plants lit up by blooms like festival lighting, to having a permanent power outage. All my blossoms dropped, which is what happens when the pollen in tomato flowers is sterilized.

As an artist, I use propaganda as a tool of empowerment. The definition is neutral; it's how you use propaganda that matters. Since propaganda communication is generally thought of as a tool of the oppressors, I say flip it and use it as a tool of empowerment. In 2017, I launched my *Growing the Food Sovereignty Revolution* poster series. They're posters about seed and food sovereignty, and they are meant to point out injustice and empower people to claim their rights, not deny them.

Local heirloom varieties also get posters as a tool of empowerment for them. They can't speak for themselves, so these posters do that for them. Their stories need to be told.

I have one for Salem County. It includes a black-and-white photo of a lot of Sparks Earlianas waiting to be shipped. The photo was taken in 1904 in Swedesboro. Burpee printed it in his *Burpee's Farm Annual for 1905* catalog. The caption reads:

> *"Shipping Sparks' Earliana Tomatoes,—(thirty carloads in one day!)—*
> *from Swedesboro, N.J. "*

I saw the actual photo at Smithsonian Gardens.

His *Burpee's Farm Annual for 1905* catalog notes:

> *"Sparks 'Earliana' seed stock source was developed from farms in'*
> *'South Jersey Home', 'Buck's County, Pa.' and a 'Western-Grown Seed'."*

He was selling hard while isolating his own version. My window into that is the story you're reading, and it's truly an honor to be this connection.

The seed trials were very interesting. There were seeds from a lot of different versions of Sparks Earliana over the five years I read, so it was very popular. I did see some comments written in entries on the small sample size of my plants.

I got to see firsthand in my garden both sides of the same story: the characteristics being sold in the catalog, and the process of developing the tomato with these characteristics in the fields for seeds to sell.

My career has been in marketing and advertising, designing marketing collateral, like seed catalogs, and now I do seed work. When I say this story chose me to tell it, this is why.

From the Smithsonian Institution, Archives of American Gardens, W. Atlee Burpee & Co. Records, *Vegetable Seed Trials*, 1903. Box 50, 51.

Here are the entries of the Sparks Earliana from pages 161 and 162 of the 1903 *Burpee Vegetable Seed Trial Journals*:

#2841 Tomato, Spark Earliana
July 30, One large plant, late. Balance typical, ripe fruits.

#2842 Tomato, Spark Earliana
Aug. 12, Typical growth, and five smooth fruits, but this row set in ditch or drain and almost drowned.

#2843 Tomato, Spark Earliana
Lewis,
July 30, Two large plants, three ripe fruits, rough and irregular.

#2844 Tomato, Spark Earliana
Lewis,
July 30, One strong plant, one ripe, fruits rather small and flat.

#2845 Tomato, Spark Earliana
Lewis, July 30
Two large plants, eight fruit ripe, one a pink fruited.

#2846 Tomato, Spark Earliana
July 30,
Poor strain, growth tall, but open leaves, curled, fruits flat and ribbed, two ripe.

#2847 Tomato, Spark Earliana
July 30, Growth too tall, and running, fruits flat and ribbed, two ripe.

 #2848 Tomato, Spark Earliana
July 30, Much better than two preceding, fairly typical, growth and fruit, seven ripe fruits, smooth and deep.

Here are three examples of how Burpee wrote in his catalogs, while trialing and selecting his version of the Sparks Earliana. The examples on the left are from the seed journals of 1903. The descriptions below are from 1901, 1906, and 1907.

Tomato,—Sparks' "Earliana." This is the earliest smooth *bright-red* Tomato of good size now in cultivation. It has been developed in Southern New Jersey, where the greatest competition exists among growers to be the first in the market. MR. GEORGE C. SPARKS, of Salem Co., has been most successful in developing his "*Earliana*," but, of course, the fruits are not equal in size or quality to those of the best varieties that are slightly later in maturing. The plants are quite hardy, with rather slender open branches and moderate growth, well set with fruits, nearly all of which ripen very early in the season. The tomatoes are deep scarlet, generally smooth, and grow in clusters of five to eight, averaging two and a half inches in diameter. Flesh deep red and of slightly acid flavor. In our Trial Grounds a specimen plant had forty-five ripened fruits at time of the first picking and thirty-five green ones of good size, and yet the whole plant could be completely hidden by an ordinary bushel basket. Per pkt. 10 cts.; ½ oz. 30 cts.; oz. 50 cts.; 2 ozs. 80 cts.; ¼ ℔ $1.50; per ℔ $5.00.

Burpee Farm Annual, Quarter Century Edition, 1901, page 31, public domain

1100 Sparks' Earliana.⊙

SPARKS' EARLIANA TOMATO.

The earliest smooth bright red Tomato of good size. The plants are compact in growth, with short close-jointed branches, setting fruits very freely in the center,—*see illustration on page 88.* This habit is so well established in our selected strain that *an entire plant may be covered with an ordinary bushel corn basket;* yet so freely are the fruits set that *each plant will produce a five-eighth bushel basket of tomatoes* during the season of about four weeks. The tomatoes are quite uniform in size, averaging three inches in diameter and from two to two and a half inches in depth; they are fleshy, slightly acid in flavor, solid, and excellent for shipping purposes.

☞ *The illustration at top of page was engraved from a photograph taken by us on July 20, 1904.* ☞ *On this day we counted on the tracks eighteen box cars and twelve refrigerator cars,—all loaded with* SPARKS' EARLIANA TOMATOES! *On the preceding day (July 19) there were shipped from Swedesboro station thirty-six cars, all loaded with this single variety of Tomatoes!* Just think of one day's shipment of *22,600 crates,* and you will have some idea of "the money-making" qualities of SPARKS' EARLIANA! **Choicest seed of our own growing in its "South Jersey Home,"**—this strain is *extra selected* and absolutely the Best *on the market to-day.* Per pkt. 10 cts.; ½ oz. 25 cts.; per oz. 40 cts.; 2 ozs. 70 cts.; ¼ lb. $1.10; per lb. $4.00, postpaid; 5 lbs. or more at $3.75 per lb. ☞ *See also* page 88.

SPARKS' EARLIANA. *From a photograph taken at Fordhook Farms of a single cluster.*

Burpee Farm Annual, 1906, page 31, public domain

1097 "Sunnybrook Special" Strain of Sparks' Earliana.⊙ In southern New Jersey where hundreds of acres are set out each season with *Sparks' Earliana,* it is natural that the farmers should each strive to get large smooth tomatoes "*first on the market.*" We had an acre in the new "*extreme-early*" strain at SUNNYBROOK the past season. From the time the plants were set out there was a *noticeable difference.*

This SPECIAL NEW STRAIN made a closer-jointed and more compact vine, set quite as freely, while the tomatoes were deeper, smoother, and ready to market at least *five days earlier!* This means large additional profits, as the first shipments always bring much the highest price. Not only does the plant bear a large cluster as the crown set, but also produces a number of smaller clusters and single fruits on the side branches.

A good stock of *Sparks' Earliana* has never contained much seed. This "SUNNYBROOK SPECIAL" STRAIN contains *considerably less,* and if the type is to be maintained, the seed will always be costly. For soils of a sandy character there is no other red Tomato that will prove so satisfactory for the early market as will this "*Sunnybrook Special*" *Sparks' Earliana!* Every gardener who is so fortunate as to obtain a few seeds this

Burpee Farm Annual, 1907, page 42, public domain

Iconic Jersey Tomatoes 57

130

KILLE #7

A GLOUCESTER COUNTY, NJ ORIGINAL HEIRLOOM TOMATO

Kille #7

Originator:	Willard Bronson Kille
County:	Gloucester
Introduced:	On his farm, Orol Ledden & Sons, Stokes Seed Company

I originally published this as a blog post on my *Plant Propaganda* blog on May 29, 2019. I've updated some of it.

Five years ago, when I started my Library Seed Bank project, I knew I was going to create the change I wanted to see in the world, and I was doing it on my terms. My stretch goal was to somehow revive a lost Jersey tomato, and I never thought that I would become part of a tomato farmer's legacy of Gloucester County-bred tomatoes.

One spring day in 2018, an email popped into my phone asking me if I had ever heard of a tomato called the Kille #7. The sender was looking for information about it because her grandfather developed it, and she wanted to share her family's legacy with her daughter in the garden. She asked me if I could help her. Of course I would.

When I found out that her grandfather's farm was in the county where I grew up, where I launched my Library Seed Bank project, and where I have a network of seed libraries, I wanted to stand up and yell, "F*cking Aye," but I was at my day job in a cubicle. So, I just stood up, and yelled it in my head.

Then, I became a man on a mission. I frantically checked my first source for tomato information, *Tatiana's TOMATObase*, a comprehensive tomato wiki. I saw the Kille #7 listed there with a seed vendor, the Sand Hill Preservation Center. Before I could jump out of my skin, I had to go to Sand Hill's website, and see if seeds were available for 2018.

I got to Sand Hill's website, and they had seeds for the 2018 season! I wanted to stand up, jump out of my skin, and once again yell "F*cking Aye!" but I was still at my cubicle. So, I just stood up, and sat right back down. And stood up again. I decided to stop doing that because I didn't want to look like a whack-a-mole.

I replied to the original email (in all caps, which was not intended) that I found seeds, and asked my contact to call me. I had typed in the wrong phone number. I was off by a digit in the area code because, well, I was trapped in a cubicle and bursting with excitement.

My contact was able to figure out my phone number and called. We chatted on the phone, and I found out that Willard Bronson Kille, the farmer who developed the Kille #7, won a national farming award in 1925. It was all in a newspaper article, and a copy would be sent to me.

The family had seed pouches with Kille #7 stamped on them, but they didn't contain seeds. Now that they had a seed source, the growing legacy could continue. It would make a great story if it ended there.

When I got the copy of the article, I saw the award was presented to him by the then-governor of New York, Franklin Delano Roosevelt. After seeing that and reading the in-depth article—it was in an old broadsheet newspaper—I had a hunch.

I'm a very curious person, and when my curiosity gets engaged, tenacity kicks in and I'm off. I love history. And tomatoes. History plus tomatoes equals my bliss.

I started with some preliminary research at the Gloucester County Historical Society. I found a 2005 newspaper article about local crop history. As I was skimming it, there it was: a shout out to Willard B. Kille, his tomato experimental farm, and his most significant tomato, the Kille #7. The article said in the 1960s, the Kille #7 was a very popular tomato locally, and somewhat nationally. I wanted to jump up and out of my skin, yelling "F*cking Aye," but now I was in a research library. So, I stood up and asked for a photocopy of the article instead.

The natural progression in my mind now was for a county proclamation. This tomato had to be declared a Gloucester County, NJ, original heirloom tomato. Is that a natural progression of thought? Well, maybe not, but when your alias is Tomato Quattrone it is.

I asked the mayor of a town that I'm active in what the procedure was for getting a proclamation about a tomato. She passed my message along to the county commissioners' office with a strong message of support detailing the community work that I do, especially with seeds.

A couple of days later, I got a voicemail from the commissioners' office asking me to call them. I called right back, and my contact there introduced herself. She gave me the backstory about how this came to her and what the procedure was for a county proclamation. Then, in a polite tone, I heard *"Jeff, we don't do proclamations for tomatoes. We do them for people, but if you can get enough history then we will see what we can do."*

Success! This was a win. They were allowing me to make my case. That's all one can ever ask for. My birthday was coming up. The historical society was going to be open on it, and what better way to celebrate it than researching the history of an accomplished farmer, his most successful tomato, and his contribution to the county where I grew up—in all for the first proclamation about a tomato in Gloucester County.

By the end of that day of research I had seventeen pages of history. Included in the research was the award that Kille received from the state of New Jersey for being an outstanding farmer. In addition to his national award, I found out he was an active citizen in his local community and in the agriculture community. He had a great story.

I scanned all of it, composed a PDF, and sent it off. My contact from the commissioners' office called to thank me for all the research, mentioned that she was impressed with the contributions that Kille made, and closed, *"but we don't do proclamations for tomatoes."*

Some time passed, and I got another voicemail: *"Jeff, please call me."* So I did, and I heard the news. They were going to issue a proclamation for Kille and his tomato! They were impressed with the history and the significance of Kille's contribution to the history of the county.

Now I was asked, *"Where is the proclamation going to be read?"* That was a very good question. In case you haven't picked it up, I was new to the proclamation game. I was trying to organize an event around tomatoes, but things were falling through. The weather here that summer was awful for tomatoes. The Kille #7 plants that I was growing in a Gloucester County community garden started out so strong, but they got hammered by the weather like the rest of the tomato crop.

There was no event for the proclamation to be read. The county offered to give me copies unread, another step outside of protocol. They were great to work with. I asked for a copy to send to the family, and a copy for myself. What I didn't know was, I had my own WHEREAS in the proclamation, and not for pushing them to issue a proclamation for a tomato. It was for my work to get this tomato to her home soil, to provide a way for the residents to grow the local pride known as the Kille #7, and to honor Willard B. Kille for his legacy. I was humbled to read that.

Gloucester County
Board of Chosen Freeholders
Proclamation

~ In Recognition Of ~
Gloucester County's Kille #7 Heirloom Tomato
Developed by Farmer Willard B. Kille, Swedesboro, NJ

WHEREAS, it is the desire of the Board of Chosen Freeholders to recognize and pay special tribute to the development of the Kille #7 tomato by Farmer Willard B. Kille, Swedesboro, NJ; and

WHEREAS, two of the major farm crops raised in the first half of the 20th century were asparagus and tomatoes; and

WHEREAS, one of the earliest tomato research farms in the county was located in Swedesboro, New Jersey. The owner of the research farm was Willard B. Kille and his most successful tomato was known as Kille #7; and

WHEREAS, Willard B. Kille was rewarded for his contributions to the Science of Agriculture and received the high honor as a "Master Farmer" with a medal presented by Governor-Elect Franklin D. Roosevelt of New York on December 30, 1928; and

WHEREAS, Willard B. Kille was awarded the Citation for "Distinguished Service to New Jersey Agriculture" in 1953. Since 1932, this award has been given to recognize and honor, those individuals who have made outstanding contributions of public service to New Jersey agriculture; and

WHEREAS, in the 1960's, the Kille #7 tomato was known as one of the finest tomatoes grown in Gloucester County and was a leading commercial variety in many parts of the country; and

WHEREAS, Jeff Quattrone of the Library Seed Bank is working with his alma mater, Stockton University, on a farming oral history project and an Ark of Taste Seed Library. Jeff found the heirloom seeds and is growing the Kille #7 tomato at the Woodbury Community Garden with the hopes of bringing this locally sourced seed to the Woodbury Seed Library for next year's tomato season.

NOW, THEREFORE, BE IT PROCLAIMED, that I, Robert M. Damminger, as Director, and on behalf of the 2018 Gloucester County Board of Chosen Freeholders, Giuseppe (Joe) Chila, Lyman Barnes, Daniel Christy, Frank J. DiMarco, James B. Jefferson and Heather Simmons **do hereby recognize and acknowledge the development of Gloucester County's Kille #7 Heirloom Tomato.**

IN WITNESS WHEREOF, the Director and Clerk have caused these presents to be executed and the seal of the County of Gloucester to be affixed this 10th day of September, 2018.

Robert M. Damminger
Freeholder Director

Giuseppe (Joe) Chila
Freeholder Deputy Director

Lyman Barnes
Freeholder

Daniel Christy
Freeholder

Frank J. DiMarco
Freeholder

James B. Jefferson
Freeholder

Heather Simmons
Freeholder

Attest:
Laurie J. Burns, Clerk of the Board

For someone who works to preserve local heirloom foods as an heirloom seed activist, having the opportunity to help a seed get back home is a gift and a dream come true. When it's in the county where you grew up, where you learned to grow tomatoes, and where you launched your seed library project, that's divine.

Through a collaboration with the three seed libraries that were up and running in March 2019, and my Library Seed Bank project, we reintroduced the Kille #7 with a controlled grow-out program. The goal was to build a local and sustainable seed supply of the tomato so Willard B. Kille's legacy would live on permanently.

To celebrate the fifth anniversary of my project on March 2, 2019, the county proclamation was read in Woodbury, along with a city proclamation from Woodbury, which is the county seat of Gloucester County. Woodbury's mayor, Jess Seagraves Floyd, whom I asked about the proclamation procedure and who got the process started, provided a proclamation from Woodbury too. Woodbury was where I did my research at the county historical society, and it is one of the places we started releasing the seeds for the growing program through the second seed library I set up.

This story is like a seed that gets planted every year and provides new fruits or flowers as content for more stories about it. In 2021, I co-organized Slow Food USA's Slow Seed Summit. We programmed opening and closing keynotes, and included Seed Stories as part of the keynotes. The day of our opening, we had a hole to fill in the seed stories, so I said I would tell the story of Kille #7, which I did.

I had recruited Phil Kaugh, who was the director of preservation at Seed Savers Exchange, to be on our programming committee. He was in the audience and heard the story of Kille #7. He immediately sent me an email commending me on the work that I did and informing me that Seed Savers had the Kille #7 in their yearbook, but not in their catalog. That was a big relief knowing this tomato was in another set of good hands.

Phil said they didn't have much information about this tomato, asked me to share what information I had with them, and introduced me to Sara Straate, Seed Savers Exchange's (SSE) seed historian. Sara replied that an SSE employee had purchased the Kille #7 from Ledden & Sons in Sewell, NJ.

So now, not only was I sharing the research I had with SSE, but the seeds they had secured were from Ledden, the location of the seed room wonderland of my youth. The fact that they got seeds from Ledden, who were farmers in Gloucester County where Willard Bronson Kille developed this tomato, meant that the lineage on the seeds was likely as authentic as it could be.

Here is some of the research about Willard Bronson Kille that I found. Amos Kirby had a local farming news column, "County Farming News," in the *Woodbury Times,* a local daily published in Woodbury. I used to read it as a kid.

This is from an article on February 25, 1967.

> Gloucester County's representative, and chairman of the Council's Agriculture Committee, is Willard B. Kille; of Swedesboro.

> Actually the· entire. conference can be credited to Mr. Kille as the talks were all centered around problems that apply to this county.

Keep in mind that Kille got his national farming award in 1928.

From a *County Farming News* article published on October 1, 1966:

> What appeared to the writer to be one of one big new developments would be the control of Fusarium and Verticillium wilts on tomatoes. One of the most valuable table tomato varieties . grown in Gloucester County is Kille No. 7, developed by Willard B. Kille, Swedesboro.

From Amos Kirby, published January 1, 1968, in a book, titled *Logan:*

> They have give credit to the Willard Bronson Kille farm now occupied by Monsanto Co., where some of the very important tomato tests were conducted by research specialists from the Campbell's Soup company, NJ and U.S Departments of Agriculture.

That book *Logan* recounts the history of Logan Township, NJ. Since that first contact in 2018, I've been in touch with members of the Kille family looking for seeds. In 2023, I was able to provide seeds to Kille's great-grandchild that I had sourced from SSE, whose seed lineage goes back to the previously mentioned purchase from Ledden & Sons when they were in Sewell, NJ. To be able to connect this family to their family's seeds is an incredible honor.

The Kille #7 is on Slow Food USA's Ark of Taste.

PROCLAMATION
BY THE MAYOR AND COUNCIL OF THE
CITY OF WOODBURY HONORING
WILLARD BRONSON KILLE
JEFF QUATRONE
& KILLE #7

WHEREAS, Woodbury, the County Seat of Gloucester County, recognizes Willard Bronson Kille as an extraordinary Gloucester County farmer; and

WHEREAS, Mr. Kille had an experimental tomato farm in Nortonville, and bred the Kille #7 tomato, recognized as the tomato of Gloucester County and somewhat nationally in the late '60s.; and

WHEREAS, Jeff Quattrone of the Library Seed Bank secured seeds from the single seed keeper with seeds, and grew the Kille #7 in the Woodbury Community Garden in 2018, the first time in decades the tomato has been grown in Gloucester County; and

WHEREAS, on this date, March 2, 2019, seeds for the Kille #7 will be released from the Woodbury Public Library's Seed Library to kick off a growing program to grow and collect seeds to ensure a steady local supply of Kellie #7 seed for the future.; and

WHEREAS, the other seed libraries in Gloucester County in Pitman and Westville will be releasing seeds on March 9 2019, and March 16 2019 respectively to participate in this program to ensure a steady local supply of Kille #7 tomato seeds; and

WHEREAS, on this date, March 2, 2019, seeds for the Kille #7 will be released from the Woodbury Public Library's Seed Library to allow for the beginning of a growing program; and

NOW THEREFORE BE IT PROCLAIMED THAT, The Mayor and Council of the City of Woodbury are extremely proud of their achievements in harvesting a locally propagated tomato and their dedication in seeing to its re-introduction to locally sourced produce for the residents of both the City of Woodbury as well as the entirety of the County of Gloucester.

Signed this second day of March, in the year of two thousand and nineteen.

JESSICA FLOYD
MAYOR

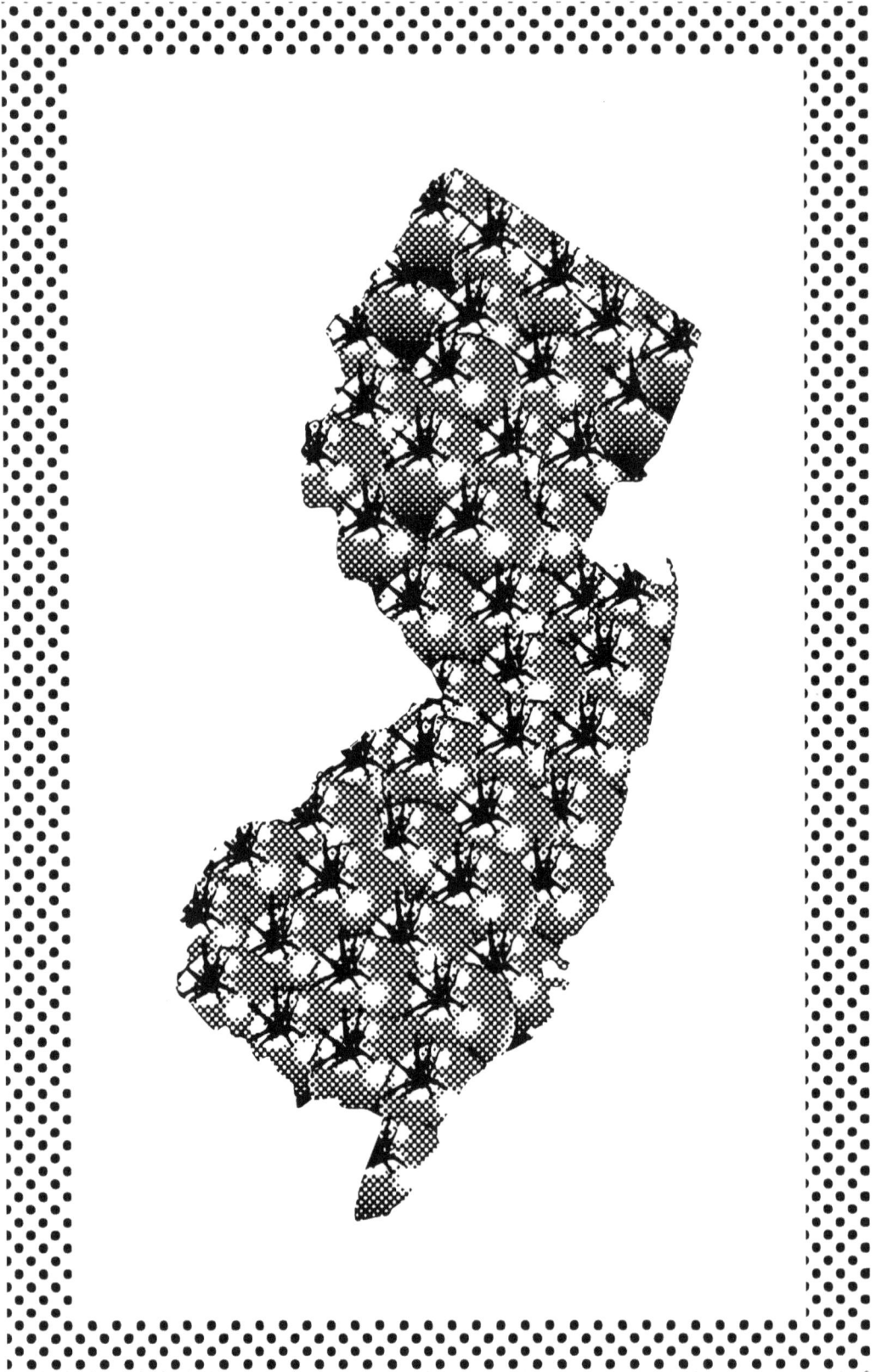

Garden State

Originator:	Campbell Soup
County:	Camden
Introduced:	1947 for their own use

That we are.

Legally now, too, since 2012, but does that really matter? It's been on our license plates since the 1950s, after the state legislature overrode a veto by Governor Robert Meyner.

Credit is given to Abraham Browning of Camden for the first reference that gave New Jersey the nickname "the Garden State." According to Alfred Heston's 1926 two-volume book *Jersey Waggon Jaunts*, Browning called New Jersey the Garden State while speaking at the Philadelphia Centennial exhibition on New Jersey Day (August 24, 1876).

Browning said that our Garden State is an immense barrel, filled with good things to eat and open at both ends, with Pennsylvanians grabbing from one end and New Yorkers from the other. The name stuck ever since.

However, Benjamin Franklin is credited with a similar comparison of New Jersey to a barrel tapped at both ends. Some have used that to discredit Browning as the originator of the Garden State.

The formality of the law really doesn't matter. Some things in Jersey just are. In this case, we know it because of our backyard gardens, our neighbors' gardens, or community spaces.

We also know our tomatoes are the best. This is a hill I will die on, as all Jersey people should.

So how does a tomato named for a state with a long tomato history, whose local pride is the tomatoes in our gardens, and which has been known as the Garden State since 1876, become functionally extinct? You might be quick to dismiss it as a *"well, it is New Jersey, so..."* —and if you did, you might have some Jersey in you.

But the truth of the matter is, it's a Campbell's Soup variety from a tomato breeding program that helped build a large multinational corporation. Campbell's is a food processing business, not a seed business, so they never released their seeds to the public through a seed company, even though large seed experimental farms were in South Jersey from Burpee, Johnson & Stokes, Wm. Maule, and P.J. Ritter, and Edgar Hurff was a noted seedman. That's a high concentration of experimental seed farms in a relatively small area.

Campbell's would release plants to farmers who grew for them. And that's how I found this tomato. When I started my Library Seed Bank project, I wanted to build a collection of old seed catalogs, since they are the seed encyclopedias of their time. I had read that statement in a seed journal once, and it stuck with me.

I was randomly searching "seed catalog" on eBay one night, and I found two seed catalogs from Ledden & Sons. I couldn't believe it! One from 1936, and one from 1955. Of course I immediately ordered them, and when they arrived, I opened them up to the tomatoes section. In the 1955 catalog, there it was, the Garden State tomato. The three parents that were used to breed it were there, encyclopedia-like, and this short description:

> New Jersey Certified Seed. 80 days maturity. Developed from crosses
> involving Pritchard, Marvel and Pink Topper. Plant vigorous, partially
> resistant to fusarium wilt and late blight. Fruits slightly flattened globe
> shape, attractive incisor and size, thick walled and mild flavor.

I went to find seeds, and there was nothing. No seeds, no history. I thought, way to go Jersey. I mean, WTF. So I searched for the three parents and I could only find two of the three, and I thought, this is a breeding program someday, the Garden State 2.0.

I put it away. In 2017, I reconnected with my alma mater Stockton University in Galloway, NJ. I met Ron Hutchinson, Associate Professor of Biology, who runs the one-acre experimental farm on campus. It's a great setup for students in the Sustainability program to learn about agroecology and market growing.

I mentioned the Garden State, and Ron said, maybe it would be a good project for a student. A couple weeks later I heard from Ron. He found the seeds at the USDA Germaplasm bank. On the USDA Germaplasm website I found out this is a Campbell's Soup variety. They donated the seeds in 1960. So, thanks to Ron and the USDA Germaplasm bank, I was able to find two functionally extinct Jersey tomatoes. I will be forever grateful.

During this time I was doing some general research, and I came across a reference for the Garden State tomato. It was in a booklet from a meeting of the Ten-Ton Tomato Club during the '50s. The Ten-Ton Tomato Club was an award program started by the New Jersey Canners Association, which Harry F. Hall, from Campbell's, had a hand in "pushing. This booklet contained a lot of information about growing tomatoes, farmers, and issues connected to farming. In it I saw that Willard Bronson Kille was quoted talking about farm labor.

The Garden State and Improved Garden State were both tomatoes grown by club award winners. One thing I didn't understand was, how did Ledden have seed to sell? In 2021, I got to interview Dale Ledden, the last member of the Ledden family, who told me his family were farmers in addition to having the farmer supply and seed business. I confirmed that his family grew for Campbell's.

That same year, the Center for Environmental Transformation (CfET) in Camden, NJ, Stockton University, and Christina Ritz, a master gardener who grows the most amazing tomatoes I've ever seen, grew the Garden State tomato. This was the third grow-out I've done with CfET using Campbell's Soup tomato varieties. Jon Compton grew the Garden State there for this project. Campbell's once made soup in Camden, but food production left in the late 1980s, and the farming was moved to California since they needed year-round growing conditions. With climate change, that's going to put pressure on that supply line.

The Garden State tomato was formerly known as Hybrid 37 or Campbell's 37. When I got tagged in a Facebook post asking if I had any information on the Campbell's 37, I didn't. I've seen it referenced, but there's not much information in public sources about the Campbell's breeding program. I said I would ask the Campbell's archivist. They came back with nothing in the archives.

Shortly after I shared the Kille # 7 research with SSE, I was interviewed by Phil Kauth and Sara Straate. The interview was for SSE's *Farm House Companion* magazine that they send to their members. The angle was how an everyday guy from Jersey doing seed preservation work with the Kille #7 demonstrates that anyone can do it. During the interview, the Ten-Ton Tomato Club came up.

After the interview, Sara searched The Internet Archive for Willard Bronson Kille and the Ten-Ton Tomato Club. She found three references from Pennsylvania. In the 1943 document, an ad for Pennheart tomato seeds mentioned Kille. I started looking through the document, and on page 53 there it was at the start of a new section: the Garden State tomato, and in parentheses, "(formerly hybrid 37)"!

Now I had the history of the Garden State tomato, and answers to the Facebook question about whether I knew anything about Campbell's Hybrid 37. From the search for a Ten-Ton Club and Willard Bronson Kille came the history of the Garden State tomato. It's the only history I could find, since there was nothing online, and the Campbell's archivist said they didn't have anything in their archive.

If Kille's granddaughter had never contacted me, there wouldn't have been a search for the Ten-Ton Tomato Club and Willard Bronson Kille. I'm surprised Ledden didn't pop in somewhere.

Here is the history of the Garden State tomato from the 1943 *PA Ten Ton Tomato Club* booklet, written by D.R. Porter, page 57. There is an asterisk by the name D.R. Porter with a footnote at the bottom of the page referencing "Research Department, Campbell Soup Company, Riverton, N.J."

> *History. Garden State was created by Mr. W. E. Belleville of the Campbell Soup Co., under the direction and encouragement of Mr . Harry F. Hall, formerly of this Company and recently retired. It resulted from crosses involving Pritchard, Marvel and Pink Topper. The early breeding work was initiated in 1931 in our greenhouse at Riverton, N. J. Significantly, none of the three varieties used in the above crosses is of value to growers of canhouse tomatoes. Pink Topper is more or less a curiosity, Marvel is almost non-existent and Pritchard is seldom used for a canhouse crop.*
>
> *However, each of these three varieties has contributed to the successful development of Garden State. This is not without by precedent – many breeders have developed valuable vegetable varieties crossing types or varieties possesing many inferior but a few ave superior been lost characters. while In the "melting pot" the inferior characters so combined the superior characters have been retained and so that the new variety has demonstrated merit.*

There's also some information about four years of tests that demonstrated an earliness different from Sparks Earliana and Bonny Best. The production peak of the Garden State was earlier than Rutgers and Marglobe. The JTD and Marglobe were crossed to create the Rutgers, which was a joint adventure between Campbell's Soup and Rutgers University.

One farmer had a yield of 20.6 tons from the 1943 test year, with twelve growers yielding more than ten tons per acre. In these tests, the Garden State also yielded bigger fruit than the Rutgers. In 1943, the Garden State yielded 145 fruits per basket, while Rutgers yielded 160.

It's an interesting snapshot of the Campbell's breeding program here. They started the early breeding work on the Garden State in 1931, while Rutgers was finishing up its field tests and selections for the Rutgers tomato. The Rutgers tomato was released in 1934.

From page 59 of the PA Ten Ton Club pamphlet:

> Seed supply for the 1944 crop is limited but adequate for small acreages with many growers. We suggest that men who have not grown Garden State plant 3 acres or less in 1944 and study it from the standpoint of fertilizer response so that they may plan for 1945.

> Stock seed has been released to several commercial Seedsmen and their rate of seed increase will probably be governed by the demand. These Seedsmen will have no seed of Garden State for sale for the 1944 crop.

I found the seed listed in the 1955 Ledden & Sons catalog. Campbell's donated the seed to the USDA Germplasm bank in 1960. I wonder if Campbell's was about to replace the Garden State with new and improved varieties that they developed at that time. If they were, it's now back in New Jersey starting in 2021.

I'm very impressed with the flavor of this tomato, the production, and the number of seeds.

The Garden State is on Slow Food USA's Ark of Taste.

Campbell's 146

Originator:	Campbell Soup
County:	Camden
Introduced:	1950s for their own use

Developed in the late '50s, the Campbell's 146 sure is a tasty tomato. The seeds are becoming widely available. While it's not one that I revived, it's a great tomato. It became Campbell's standard for juice flavor. It does have a great flavor, and it was part of the first Campbell's Soup tomato-growing program I did with CfET.

The focus of that first-grow out was to go as far back in the history of Campbell's tomato breeding program as possible. We grew the JTD, the first tomato bred by Campbell's and released in 1910, the Marglobe, and the Rutgers. The Marglobe is not a Campbell's variety, however it was crossed with the JTD to breed the Rutgers tomato. These three tomatoes captured a significant part of Jersey tomato history.

The Campbell's 146 represented the continuation of Campbell's tomato breeding program, and an opportunity to connect this breeding program to impacts on American pop culture and American food culture because of the growth of Campbell's brand, and changes in American society and the food system.

One impact is on food culture with the iconic American comfort food combination of tomato soup and grilled cheese sandwiches.

Cheese and bread as a combination has been around for a very long time. Toasting it in a sandwich is up for debate; the 1920s is a common timestamp in the history I've found in my research. That said, the French were making their croque monsieurs since the early 1900s. Grilled cheese sandwiches were originally called toasted cheese or melted cheese sandwiches. They were open-faced with grated American cheese. That all started to change when Kraft introduced Kraft singles in 1949. It wasn't until the mid-1960s that the term grilled cheese appeared.

In 1914, James Kraft opened his first plant for his patented processed cheese. The invention of a bread slicer and wrapper by Otto Frederick Rohwedder in 1927 paved the way for sliced bread to become commercially available at an industrial scale. Wonder Bread was introduced in 1930, and its success demonstrated to other companies how they could profit from sliced and packaged bread, which explains the bread aisle in supermarkets today.

Toasted cheese sandwiches became popular with the American public during the Great Depression. During WWII, the government issued cookbooks that Navy cooks used to prepare American cheese sandwiches. Sliced and packaged bread and processed cheese that could travel without immediate spoiling came together, and met the nutritional standards of the Navy.

After the war, institutional food service standards were updated, including for school cafeterias, and it was decided that another component was needed to round out a meal with grilled cheese sandwiches, so tomato soup was added.

Condensed Campbell's tomato soup in a can, sliced and packaged bread, and processed cheese that didn't immediately spoil were all products of a growing industrial food system. They're cost effective and shelf stable.

For the American consumer, these three foods were economical, and they taste good together. It's an interesting timeline, one that reflects the move away from local seed and food sovereignty.

Campbell's straddled the line with their tomato breeding program. They bred seeds for locally strong, productive plants that they released to local contract farmers. This provided food to the local community, while the intention was to manufacture convenient industrial food. The seeds weren't released, so they retained sovereignty over them.

The Jersey tomato was there from the start with Campbell's. As Campbell's grew, it became an iconic soup brand in large part because of Jersey tomatoes. The can's label never changed much. In 2021, the labels were refreshed with the first design changes in fifty years.

It's thought that this consistent image branding was one of the reasons that Andy Warhol chose Campbell's Soup cans for his *Soup Can Series*. Warhol had a memory of eating Campbell's Soup while growing up.

From the history.com website:

> *"I used to drink it," he said. "I used to have the same lunch every day for twenty years."*

Food memories are powerful, and in this case, potentially inspirational in a series of paintings that became part of the Pop Art movement.

Warhol's show opened on July 9, 1962, at the Ferus Gallery in Los Angeles. At first it wasn't well received, but innovators are ahead of the curve and it takes time for the public to catch up. Eventually, critics and the public started getting what Warhol was doing. Warhol switched from painting to screen printing so he could create more pieces, and his stature rose.

Campbell's always used a blend of tomatoes. The Campbell's 146 tomato was likely part of the blend of tomato varieties in the cans Warhol painted. When Warhol painted his series, Campbell's had 32 varieties of condensed soup. He painted every soup variety. The tomato soup can is the most iconic in the series.

In every can of Campbell's soup for a very long time, there were Jersey tomatoes. Our Jersey terroir gave birth to the Sparks Earliana and all the varieties of tomatoes that followed. The Campbell's breeding program. The experimental farms of Burpee, Johnson & Stokes, (later the Stokes Seed Company), and the Wm. Maule Seed Company all played a role. Willard Bronson Kille, an award-winning farmer had his own tomato experimental farm. The Ritter Family of Bridgeton, developed their own variety of tomato for their ketchup in their specialty foods canning company, P.J. Ritter, and had a seed company division that in the '50s claimed to be the largest supplier of certified tomato seed in the world.

Whether it was for a local brand like P.J. Ritter, or the fruit that blossomed into a large multinational food manufacturer like Campbell's, it was Jersey tomatoes.

As part of an iconic and beloved American comfort food combination or an innovative Pop Art show that caught or generated the wave of the Pop Art Movement, Jersey tomatoes are there.

Across this wide swath of social and cultural elements, the local pride that we grow in our soil, the Jersey tomato, shows up, like Jersey always does. And the seeds of the Campbel's 146 connect us to all of this.

Stokesdale and Valiant

Originator:	Stokes Seed Company
County:	Burlington
Introduced:	Stokesdale 1936, Valiant 1937

The Stokesdale tomato is emblematic of the history of the Stokes Seed Company when they were in Vincentown. Introduced in 1936, the Stokesdale tomato became the marquee tomato for this seed company, whose Jersey history goes back to 1878, and played an important role in Jersey tomato history.

When Stokes introduced this tomato, they made a bold change in their business model. They switched from selling vegetable, flower, and herb seeds to just selling tomato seeds. Kind of an epic move.

Stokes bred the Valiant tomato from the Stokesdale, and both played a prominent role in the direction and marketing of the Stokes Seed Company. In addition to featuring both, Stokes also sold hybrid versions of the Stokesdale, and they sold a few other tomato varieties that fit their new business model, including the Kille #7.

The Stokesdale tomato, in my opinion, is a blue-ribbon tomato. Its flavor is fruitier than the other Jersey-bred tomatoes I've tasted, and slightly sweeter, but still has the distinct acidity that most Jersey tomatoes have. It's slightly larger and denser.

The Valiant, so far in very limited field trails, hasn't been as impressive as the Stokesdale. The research shows that it's a tomato suited for juicing. That makes sense since there was a tomato juice canning factory on the property of the Stokes Seed Company.

I haven't found a documented connection showing that the Valiant was bred exclusively for juice. The seeds for this variety are very rare, and that could be a factor in the under performance so far. I am confident though that with proper selection, this tomato can showcase the characteristics that inspired the Stokes Seed Company to market it so prominently.

Both were revived in 2022, with the seeds for the Stokesdale tomato coming from the USDA, and the seeds for the Valiant tomato coming from a preservation farm. The Stokesdale is on Slow Food USA's Ark of Taste.

Stokesdale —

NEW LAST YEAR, EARLIER AND LARGER THAN MARGLOBE

STOKESDALE is proving worthy of its name. Although now only in its fifth generation and requiring at least one more year for final fixing of type, it already has demonstrated its remarkable strength. It will mature with the Bonny Best class, and that means a week to ten days ahead of Marglobe. In size will range two to three ounces larger than Marglobe. It is comparatively free from stem-end crack. What slight cracks do develop are of the concentric type. Its production is remarkable. Like Rutgers, it ripens from the inside out which is helpful on the Government Grading platform. Its rare flavor is a distinct asset.

Because of the unusual heat of July, 1936, we do not have a final judgment on its vine coverage. It did not stand the 145° field temperatures as well as Marglobe. As a result, our Proving-Ground selections have been made toward slightly heavier foliage. (Obviously too vegetative a growth means much later maturity.) At the U. S. Department of Agriculture Plant Breeding Station at Beltsville, Md., and in a test-plot in the Northern Neck of Virginia, conducted by the University of Virginia, Stokesdale did not survive in the soils which were heavily infected with Fusarium Wilt. Otherwise, various experiment stations report disease-resistant factors about equal to Marglobe.

Stokesdale is enthusiastically recommended to growers who can profit either by a quick-maturing Tomato, or who, because of high latitude or altitude, must have a Tomato that will ripen in a short season. This, obviously, makes Stokesdale important in our northern-tier states and in Canada, where full crops of Stokes Master Marglobe usually cannot develop. The St. Catharines, Ontario, Trial-Grounds of Stokes Seeds Ltd. gave a convincing demonstration of the importance of Stokesdale for the North. Favorable reports have also come to us from as far South as Florida and Texas. If you did not plant Stokesdale in 1936, we urge that you give it consideration this year. Obviously, no Tomato is suitable to all conditions, but this has made a lot of money for some of our customers. **Price, Postpaid:** Trade pkt. 25 cts.; oz. $1; 1/4 lb. $2.75; lb. $10.

Stokesdale—1938 Type

Proving-Ground stock for the experts, the premium seekers, and the breeders. This gives you the 1938 model one year in advance. It is one of the most distinguished Tomato types we have ever produced. **Price, Postpaid:** Trade pkt. $1; 1/4 oz. $1.50; 1/2 oz. $2.75; oz. $5.

[7]

Valiant —

A NEW, VERY EARLY, BUT LIGHT-FOLIAGED VARIETY

VALIANT is a word for courage. When you grow this Tomato you will understand why we have given it that name. It is indeed truly courageous. How such a small plant can produce Tomatoes of its size and season is incredible. To us the word valiant is one of the most beautiful in our language. Here again the Tomato matches this beauty. It is extremely well formed; in fact, it approaches Stokes Master Marglobe in perfection and should make a good Tomato for the Green-Wrap Trade.

In season it will mature three days after Earliana and five days ahead of Stokesdale. Therein lies its importance.

The plant of Valiant is unusually restricted. In this particular it approaches Earliana. The leaf is about one-half the size of that of Stokes Master Marglobe. *(We were tempted to call it "Naked Indian" after a tree we once saw in the Tropics.)* Take careful note of this sparseness of foliage. Do not mature it during the extreme heat of midsummer. For winter and spring production in the Far South, for early season production in the North, and for greenhouse production *any time* we strongly recommend it.

Our seed supply is seriously restricted. Probably it will not hold out for the season. If the above description fits in with any part of your production plans, come to us early in the year and give it a fair trial. Keep this in mind: Valiant is a smooth, handsome, modern-type Tomato competing seriously with the main Earliana harvest. Its direct relation to Stokesdale indicates an unusually solid interior.

Valiant will open your season brilliantly, profitably, and on time. You will like it immensely.

Price, Postpaid: Trade pkt. $1; ¼oz. $1.75; ½oz. $3; oz. $5; ¼lb. $17.50

SUMMARY FOR VALIANT: ↑

Days to maturity at Stokesdale: 107.
Germination: 95%.
Date of test: December, 1936.
Disinfection: Mercuric Chloride.
Ratio of depth to width: 90%.
Interior: Very solid.
Color: Intense scarlet.
Average weight of fruit: 6 ounces.

SUMMARY FOR STOKESDALE: ⟶⟶

Days to maturity at Stokesdale: 112.
Germination: 96% or better.
Date of test: December, 1936.
Disinfection: Mercuric Chloride, 1 to 2,000.
Ratio of depth to width: 80%.
Interior: Remarkably solid.
Color: Intense scarlet. Ripens from inside out.
Average weight of fruit: 8 ounces.

[6]

Tomatoes by Stokes, Designed for Selling, 1937, page 7, public domain

Jersey Devil and Jersey Giant

Originator:	Martin Sloan
County:	Middlesex
Introduced:	1970s by the Tomato Seed Company

One of the many gifts that this story has given me is that I get to set the record straight. In this case about a New Jersey paste tomato, one that shares the name of our beloved resident devil, and the other that shares its name with a breed of chicken.

There are two stories attributed to these tomatoes. The first is that they were released by the Tomato Seed Company of Metuchen, NJ, in 1987, and the other that they were a mainstay of the tomato canning industry.

Until I grew the Jersey Devil, I believed it was a mainstay of the canning industry since it's a paste type. It made sense. Once I grew it, I knew it couldn't be a canning tomato with its beastlike plants and independent character. The reward of letting these plants be is an abundance of very large, very sweet tomatoes. I've made ketchup with the Jersey Devil without sugar — that's how sweet they are.

The true story is that the Jersey Devil and Jersey Giant tomatoes were developed by Martin Sloan of Metuchen. They were released by his Tomato Seed Company in the early '70s, not 1987.

Martin told me he was working at a pharmaceutical company, and one of his coworkers brought in these odd-looking tomatoes from one plant in her garden. She had gotten seeds from a local drugstore. His coworker thought they might be poisonous. She knew Martin grew tomatoes, so she gave them to him.

Martin knew what was going on—a genetic mutation—and said the tomatoes were fine. He tasted them and was impressed with their sweetness, so he saved the seeds and made the selections for the Jersey Devil and the Jersey Giant. He released the seeds through his seed company, the Tomato Seed Company.

Martin told me he turned his backyard into a seed farm, and sourced seeds for over 300 varieties of tomatoes from around the world through the mail. This was before the internet. He hand-lettered his catalogs instead of using a typesetter, and they were printed by a local printer.

I have a PDF of one of his catalogs, and it is beautiful. I design page layouts, and this catalog is exquisite with its hand lettering. I've only set type electronically, never hand-lettered. My career wouldn't have happened if I had to.

These tomatoes are big and bulbous with a point at the end like a pepper, or they're narrow and long with a slight taper to the point. Some are curved at the end like a horn. The biggest one I've had was close to 9 oz. For a paste tomato, that's huge.

What led me to track down the true story? A conversation on Earth Day in 2023 with Damon Smith of Our Gardens, a community program dedicated to peace and healing through seeds and food biodiversity, at Reed's Farm in Egg Harbor Township, NJ. Reed's is home to A Meaningful Purpose, a nonprofit that addresses the needs of food insecurity, soil regeneration, and community inclusiveness in Atlantic County, NJ. Reed's was the first Slow Food Farm in the United States.

Damon told me he wanted to take the San Marzano down as the local favorite paste type with Jersey's native paste tomato. Of course, I agreed.

A true San Marzano is from the terroir that created it, in the Campania region of Italy. Everything else is a San Marzano variety. Since Jersey has its own paste tomato that was developed in our soil, and is a fantastic sweet, meaty, large tomato, it should be presented to the public so they can decide for themselves if it's better than the San Marzano.

At this point, I needed plants, and I knew one source: Pine Barrens Post, a Piney homestead with great products. I messaged them, and they saved the rest of their plants so there would be tomatoes and seeds to get started on this. Thanks to Laura Vitagliano from Pine Barrens Post for the plants. Next was finding someone to grow them, and Rob Williams took care of that.

Since this community project was taking shape, I would have to get to the bottom of the discrepancy in the history. I did with the help of a couple of public records searches. I was able to connect to Martin through his son Kevin.

Martin told me that when he was growing the Jersey Devil in the '70s, he had people in NYC working with chefs to compare it to the San Marzano tomato.

That tracks with what Damon suggested, and the first community action we took that resulted in a collection of local Jersey Devil seeds that were grown in the

Pines, the home of the Jersey Devil. I distributed a lot of the seeds at Lines on the Pines, a large cultural celebration of Piney culture, and a grow-out program with Reed's, who did a side-by-side comparison with San Marzanos.

With this grow out, they were considering replacing the San Marzanos in the sauce they make for the pizza and other products they sell. They also were able to sell some at their farm market. They were impressed with the results.

It was good to hear from Martin how in the '70s they were challenging San Marzanos for flavor, and now we are doing the same because the seeds of these tomatoes allow us to do so.

I shared some of the Jersey Devils from that grow-out with Melissa McGrath, the chef at Sweet Amelia's Kitchen and Market, Newfield, NJ, a seasonal farm-to-table market that includes a sustainable oyster farm. Melissa and her food has consistently been recognized. In 2024 Sweet Amelia's was on the *New York Times* Top 50 Restaurants in the country. In 2023, it was a semifinalist for a James Beard award. In 2022 it was a top 40 new restaurant from *Esquire* magazine, and it has been consistently recognized by the Philadelphia, PA food media since they opened in 2021.

Melissa posted on Instagram about her good experience with the tomatoes, and was able to post Martin's true story for the first time that Martin was aware of.

I'm very impressed with them. They are one of the sweetest tomatoes I've ever grown, and I've grown over 300 varieties.

With this activity, the true story of the tomato is told, as it should be. The Jersey Devil and the Jersey Giant are future nominations for inclusion in Slow Food USA's Ark of Taste.

The Bratka's Five

Originator:	Joe Bratka
County:	Bergen
Introduced:	1990s by Carolyn Male

When you start doing seed work, and you get deeper into the stories about them, you find these storied collections of seeds. The Bratka Five is one of them. As far as I know, the Bratka Five is not an official title; it's a title I've coined for them.

The story is this: Joe Bratka found tomato seeds in his father's tool shed, all with different names on them. He sent them to Carolyn J. Male, a noted tomato scholar and author of *100 Heirloom Tomatoes for the American Garden.* She got five varieties to germinate: Pasture, Mule Team, Red Barn, Boxcar Willie, and Great Divide. She released these seeds to SSE and Tomato Growers Supply.

In the story, Joe Bratka's father's name is not given. Some of the research I've found indicates that Joe was the father's name as well. Hence my title, The Bratka Five, after the family name.

The location of this tool shed was Elmwood Park, NJ, so these are a nice collection of Jersey-bred tomato varieties. Pasture is a cherry, Mule Team is a slicer, and the other three are beefsteaks with distinct characteristics.

It's especially nice to see a cherry as part of Jersey tomato history, because cherry types aren't as common in Jersey-tomato history. Beefsteaks and canning types are the most common types of Jersey tomatoes. Joe Bratka released some cherry tomato varieties he bred.

Elmwood Park is in the northeast portion of the state, a good distance away from the tomato breeding and seed production hot spots in South Jersey. I'm thrilled to see tomatoes that were bred outside of that concentrated zone. It's the Jersey tomato; it should have state-wide independent breeding.

I did have the opportunity to grow all five together in 2023. I've been a longtime fan of Boxcar Willie, and it was a goal of mine to eventually grow all five. In 2022, I grew Red Barn, and that beefsteak just hit my heart and soul in such a special way. Beefsteaks are my favorite tomato type, and Red Barn just hit every note of what a true Jersey beefsteak is. What I love about Jersey beefsteaks is the size and shape of them. When you slice them you get a thick juicy slice of a meaty and juicy tomato. When I see a slice I know I'm about to enjoy a tomato sandwich full of that tomato flavor, with the juice making the toasted bread soggy and dripping on the plate to sop with the crust of the sandwich bread. That rite of summer makes tomato lovers rejoice.

After that I knew I had to grow them all. I had Boxcar Willie, Red Barn, and Mule Team seeds. I had the Mule Team seeds from 2022 when I got the Red Barn seed. I just had to get Pasture and Great Divide. Easier said than done. Pasture I got through the Exchange at SSE, and Great Divide, the first set of seeds had zero germination, and the sources for this tomato are few. I did get seeds for 2023 from a 2022 grow-out, so I was psyched, but the seed sowing season by this point was moving on.

I direct-sowed the Great Divide in the first week of June. In South Jersey that's when the soil is the proper temperature for germination, and the cut off for direct sowing to allow time for a harvest.

My seeds for these and a couple other varieties germinated and grew. The Great Divides were growing well, with a manageable plant height and heavy big fruit set, and I was psyched. But my plot flooded from torrential rains, including 8.5 inches in a three-hour stretch overnight. The plants and tomatoes held on until the last flood. They toppled over finally on the third flood, and that did them in. I was able to harvest seeds from one tomato, and will be looking into heat treatment for them, and the other seeds I harvest.

In 2024, the Great Divide was grown at Reed's Farm in Egg Harbor Township as part of an education series about New Jersey heirloom vegetables. It got rave reviews for flavor, and will likely become part of the local seasonal Jersey tomato cuisine.

It was fascinating to see five different varieties bred by the same breeder. I'm looking forward to getting these all well-established here in Jersey so we have ample seeds for all five.

Atlantic Prize

Originator:	Unknown Atlantic County NJ, tomato farmer
County:	Atlantic County
Introduced:	1889 by the Johnson & Stokes Seed Company

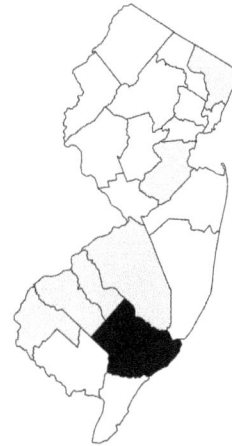

This was the first tomato introduced by Johnson & Stokes in 1889. It was selected by a farmer in Atlantic County. Atlantic County doesn't have the rich, fertile soil of New Jersey's Coastal Plain that was so crucial in the development and breeding of Jersey tomatoes. Instead, it has the sandy, acidic soil that supports the Pines, and is noted for its blueberry production, not tomatoes. Hammonton, in Atlantic County, claims to be the blueberry capital of the world.

With its rare Atlantic County origin, the Atlantic Prize has a noted place in Jersey tomato history.

I couldn't find seeds in the U.S., so I had to search internationally. I found seeds in Tasmania, South Africa, and Belgium. I contacted the grower in Belgium in 2022, and he had one package of seeds from his most recent-grow out. Who's to say when he would grow them out next, and when seeds would be available again.

I traded seeds and got fifteen to start. Five went to Liam Duffy from Homebody Farms, Chesterfield, NJ, to grow. The tomato is small by current standards. The plants are prolific and produce a tomato that works well in salads and canning.

This tomato has a special place in my heart. It represents the pride, joy, and dedication of an anonymous Atlantic County farmer, one who found it so special that they standardized the variety, included the name of the county where it originated and the word "Prize" so it could tell its story to future generations who have the seeds for this tomato.

Finding the seeds overseas shows that seeds migrate and some have a migration story of survival. Now that they're back, they're part of food insecurity growing programs in Atlantic County, and the flavor has sparked culinary interest too.

Seed and Food Sovereignty

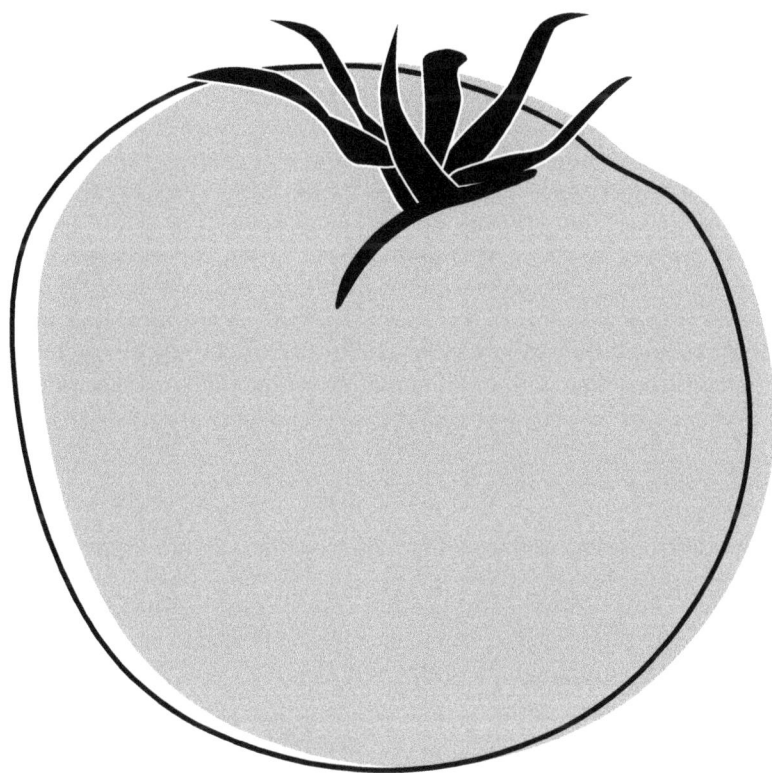

Declaration of the Forum for Food Sovereignty, Nyéléni 2007

We, more than 500 representatives from more than 80 countries, of organizations of peasants/family farmers, artisanal fisherfolk, indigenous peoples, landless peoples, rural workers, migrants, pastoralists, forest communities, women, youth, consumers and environmental and urban movements have gathered together in the village of Nyéléni in Sélingué, Mali to strengthen a global movement for food sovereignty. We are doing this, brick by brick, as we live here in huts constructed by hand in the local tradition, and eat food that is produced and prepared by the Sélingué community. We give our collective endeavor the name "Nyléni" as a tribute to and inspiration from a legendary Malian peasant woman who farmed and fed her peoples well.

Most of us are food producers and are ready, able and willing to feed all the world's peoples. Our heritage as food producers is critical to the future of humanity. This is specially so in the case of women and indigenous peoples who are historical creators of knowledge about food and agriculture and are devalued. But this heritage and our capacities to produce healthy, good and abundant food are being threatened and undermined by neo-liberalism and global capitalism. Food sovereignty gives us the hope and power to preserve, recover and build on our food producing knowledge and capacity.

Food sovereignty is the right of peoples to healthy and culturally appropriate food produced through ecologically sound and sustainable methods, and their right to define their own food and agriculture systems. It puts the aspirations and needs of those who produce, distribute and consume food at the heart of food systems and policies rather than the demands of markets and corporations. It defends the interests and inclusion of the next generation. It offers a strategy to resist and dismantle the current corporate trade and food regime, and directions for food, farming, pastoral and fisheries systems determined by local producers and users. Food sovereignty prioritises local and national economies and markets and empowers peasant and family farmer-driven agriculture, artisanal - fishing, pastoralist-led grazing, and food production, distribution and consumption based on environmental, social and economic sustainability. Food sovereignty promotes transparent trade that guarantees just incomes to all peoples as well as the rights of consumers to control their food and nutrition. It ensures that the rights to use and manage lands, territories, waters, seeds, livestock and biodiversity are in the hands of those of us who produce food. Food sovereignty implies new social relations free of oppression and inequality between men and women, peoples, racial groups, social and economic classes and generations.

In Nyéléni, through numerous debates and interactions, we are deepening our collective understanding of food sovereignty and learning about the realities of the struggles of our respective movements to retain autonomy and regain our powers. We now understand better the tools we need to build our movement and advance our collective vision.

★ MY FOOD ✊ MY RIGHT

"Food sovereignty is the right of peoples to healthy and culturally appropriate food produced through ecologically sound and sustainable methods, and their right to define their own food and agriculture systems. It puts the aspirations and needs of those who produce, distribute and consume food at the heart of food systems and policies rather than the demands of markets and corporations."

– Declaration of Nyéléni, the first global forum on food sovereignty, Mali, 2007

In my art and activism, the focus of my work is empowering seed and food sovereignty. It's through this lens that I feel confident writing this book and positioning the Jersey tomato as an American folklife and foodway. I'm looking at the tomato breeding history, and the local food history and culture of Jersey, specifically South Jersey, where Campbell's breeding program and independent experimental seed farms were. Seed and food sovereignty existed here; no one know about it or called it that because there was no need for these social actions.

The two world wars scaled up production capabilities and technology. After World War II, there was a massive social shift away from the local, which included access to personal transportation devices known as cars. The food system moved right along with it. In 2024, we're at a place where four agricultural chemical companies own up to 70% of seed lines, and ten food companies control most of the brands that are on supermarket shelves.

Growing up, I experienced what food sovereignty was like, and like everyone else at the time, I didn't know what it was. It was a ritual of growing and preserving food passed down from my migrant grandparents to their offspring in this new land far away from their local foods and culture. Like all migrants, my grandparents left a lot of their local food culture behind.

One ritual was stringing hot and sweet peppers up and drying them. My maternal grandmother dried them in her basement. I continue drying them to this day, but I use a dehydrator.

The definition for seed sovereignty is taken from the *Lexicon of Food*:

> *Seed Sovereignty reclaims seeds and Biodiversity as commons and public good. The farmer's rights to breed and exchange diverse Open Source Seeds which can be saved and which are not patented, genetically modified, owned or controlled by emerging seed giants*

The *Open Source Seeds* page from the Open Source Seed Initiative website says that:

> *Inspired by the free and open source software movement that has provided alternatives to proprietary software, OSSI was created to free the seed – to make sure that the genes in at least some seed can never be locked away from use by intellectual property rights.*

This is a great initiative. I encourage everyone to support it.

While the above definition is specific to farmers' rights, the right to save seeds is a universal human right. It's my primary motivation for establishing seed libraries.

The Green Revolution was launched in 1941 with a western savior notion of noble purpose, addressing the fear of famine in the world. It laid the foundation for large-scale industrial agriculture and food business, and the globalization of food. It also provided a great propaganda opportunity to hide the globalization and consolidation of multinational industrial agriculture and food agendas. Like most projects, there was success and failure.

The Green Revolution used the technology of hybrid seeds, agricultural chemicals, irrigation, and cultivation mechanics to increase the number of calories per acre and create new markets and profits for the companies selling the technology involved. These technologies and practices were seen as a way to supplant traditional, more labor-intensive methods. The new technology cost money, which exploited any existing farm income inequality, pushing out small farms who couldn't pay to play in the new farm technology market.

The Green Revolution was supported by the United States government, the Ford Foundation, the Rockefeller Foundation, the United Nations, and the Food and Agriculture Organization, an agency of the UN specializing in hunger. That's a lot of power and money that failed to include local farmers as stakeholders.

It did increase calories per acre, and Norman Borlaug, the father of the Green Revolution, won a Nobel Peace Prize in 1970. He's often credited with saving a billion people from starvation, but there's no documentation to support this. There are statistics about increasing crop yields with higher caloric content per acre, but no population documentation explaining how the number of a billion was derived.

The fear of famine in the developing world drove the implementation of the Green Revolution. Hunger still exists, and today, we waste a lot of food. Food waste has grown since globalization.

Supplanting the local food producers with the industrial model, however, is what created the need for "food sovereignty."

La Via Campesina, or the International Peasants' Movement, originated the term "food sovereignty" in 1993. When they started, they were a group of women and men from four continents. They wanted a voice in their day-to-day existence, which was being challenged by the globalization of agriculture policies and agribusinesses.

La Via Campesina promotes a Universal Declaration on the rights of peasants and other people working in rural areas, which includes the right to life and adequate standards of living, the right to land and territory, and the right to seeds, information, justice, and equality between women and men.

In February 2007, the first international gathering of delegates attended the Forum for Food Sovereignty in Sélingué, Mali. About 500 delegates from more than eighty countries adopted the "Declaration of Nyéléni." This chapter opens with a quote from that declaration.

In April 2008, the International Assessment of Agricultural Knowledge, Science and Technology for Development (IAASTD), an intergovernmental panel sponsored by the UN and World Bank, came up with this definition:

> *Food sovereignty is defined as the right of peoples and sovereign states*
> *to democratically determine their own agricultural and food policies.*

In September 2008, Ecuador included food sovereignty in its constitution. In 2010, the US Food Sovereignty Alliance (USFSA) was formed. In 2011, 400 people from 34 European countries met to start a food sovereignty movement there.

The need to claim food rights is relatively new, and when you look at the history of farming and food production in South Jersey, you can see the complete process. Seed and food sovereignty started with seed experimental farms, locally bred and adapted seeds, local seed companies, and food production and canning companies.

The farmers grew what the canners canned from seed bred on the experimental farms. The canners and seedsmen in some cases were one and the same. With the fishermen, crabbers, clammers, and oyster farmers, it was from the water to the market or the can.

The process and supply chain were all under local control, and complete. Because of this, there was no need to consider food as a human right. The power and control over food was local.

Remove the local, and the power of sovereignty is taken away.

Iconic Jersey Tomatoes **95**

Jersey's Foodway of Seed and Food Sovereignty

LAND of PLENTY

ACCORDING TO FEEDING AMERICA, IN 2015, THE USA FOOD INSECURITY RATE WAS 13.4%

This photograph is from 1940 titled: Distributing surplus commodities

Food Insecurity 2017 © Jeff Quattrone 2025

Jersey gets knocked a lot. I know I knock it when it's appropriate. That's the thing with Jersey culture: we never leave you guessing. We don't have time for that. We have too much to enjoy here. As Jersey native Anthony Bourdain said, "*To know Jersey is to love her.*"

The Jersey peninsula has 1,792 shoreline miles, according to the NOAA Office for Coastal Management. Along these shores, surrounded by three different water types, we have five soil types, including some of the best soil for growing crops in the country, and the sandy soil that supports the unique bio diverse Pines.

It's no wonder we're the Garden State. That barrel, the description used by Abraham Browning who first called us the Garden State, brought goods to New York City to the north and Philly to the south. Fin fish, shellfish, and muskrats from the waters. Yes, muskrats. In Salem County, locals would trap muskrats in the winter to sell the pelts and meat, and in the summer they'd trap crabs from the same waters. For hunters, we have wild turkeys, rabbits, and deer, and for foragers, woodlands, pastures, and Pines.

And yet, within this spectrum of foods from land and water, food insecurity is real. It's a paradox that needs to be part of my conversation when talking about foodways, especially in a state with abundant resources for food production. A state where tomato seed breeding, seed experimental farms, and canning provided robust seed and food sovereignty as part of the fabric of life and Jersey culture. Systematic change is needed here, and everywhere, to address insecurity.

Hunger has always been a part of society, and yet we tamed wild plants with selection and breeding that became agriculture to secure seeds. We domesticated livestock for food as we evolved away from hunting and gathering. We have the means, but not the desire to change food insecurity.

There's a lot to debate about why. As a seed and food activist, the complete story has to be told, and food insecurity will be part of the story I will tell.

My life in South Jersey began as the huge social shift after WWII was ramping up. With newly minted suburbs being converted from farmland, cars became necessary as the population started moving from urban areas to suburbs.

Retail business consolidated from local to regional and national markets. Television offered a new national advertising medium that could visually sell the modern conveniences taking hold with advances in manufacturing, especially with food. TV dinners became a thing, a product. Radio dinners as a product never existed.

The towns that were so dependent on the local seed, farming, and canning businesses became vestiges of the local economic communities. Trying to recalibrate in a global service economy, the struggle for redevelopment is real.

When you can feed yourself, your family, and your community, there's security in that. With security comes pride, which is the core of the Jersey culture. Some would say it's attitude; we embrace that. You see the key to attitude is to have it, and let others acknowledge it. It's a quiet way of boasting.

As prideful as we are here, food insecurity is nothing to be proud of. With this project, I will include food insecurity. I will collaborate with frontline activists. I will give seeds away. I will bring this food insecurity need to the public and encourage folks to create the systemic change to secure food for all.

Food security starts with good soil. We covered Jersey soil types earlier, but it's often overlooked as part of foodways. It's not fabulous enough for Instagram, but try having food without it. It's often interchanged with "dirt" which can include soil, but also includes other types of matter. Soil is a complex ecosystem that gets abused by lack of cover crops, strip mining, erosion, and chemicals, including agricultural ones. It needs to be nurtured.

It's important to note, the quality of soil is so good here that seed companies had experimental farms here. It's a good question to debate how extensive the Campbell's Soup tomato breeding program would have been without our Coastal Plain soil.

Tomatoes grow from seeds. Tomato seeds were bred to thrive in the local soil and climate. Nature created seeds to germinate in the intersection of soil and climate. Seed breeding, specifically tomato seed breeding in South Jersey, thrived in part due to the great soil we have here and is an important part of the Jersey way of seed and food sovereignty.

Soil is the first part of Jersey's historical seed and food sovereignty trifecta: soil, seeds, and canning.

Let's talk seeds. My Library Seed Bank project that launched in 2014 really started in January 2013. I studied seeds, the seed business, seed libraries, and scratched the surface of the seed history here. While I have been growing most of my life, my experience with seeds had been nil until then.

I didn't fully appreciate the relationship between seed and food sovereignty until a couple of years in. I knew it on a surface level, and as an activist, I see the same thing. Just look at policy. Seeds and biodiversity are not a major focus when it comes to policy, and policy drives action.

The reality, though, is that seeds are food. We wouldn't have a lot of it without them. They are a bit more abstract than food itself. A red cherry tomato seed doesn't look any different from a bicolor beefsteak seed, yet the genes inside tell their story in texture, taste, color, and disease resistance. That's biodiversity. Same with pepper seeds. A bell pepper seed looks similar to a Carolina Reaper seed, yet try rubbing your eye after handling both, and you will feel the difference.

Seeds are power. If you have seeds and no place to grow them, you're in a stronger position than someone with land and no seeds. You can always improvise a growing area, but you can't improvise seeds.

Once you get into seeds, and seed breeding, this becomes very clear. And here in South Jersey in the late 1890s, a scientist who figured out how to remove water from soup to pack it in cans instead of glass jars, thereby reducing shipping costs understood this. Campbell Soup Company bet big and early on agricultural research. While you might not think about the tomato in their tomato soup can, they did. They bred their own and built a large multinational food business based on their tomato seed breeding. John Thompson Dorrance was that scientist.

Campbell's is one example of a canning company here breeding their own tomato. The seed and canning business model existed here with much success. There were large experimental farms here from the Johnson & Stokes Seed Company, Burpee had their Sunnybrook Farm here, and so did Wm. Maule.

Willard Bronson Kille, the farmer who bred the Kille #7, sold his own tomato seeds. Orol Ledden & Sons had a seed business here. They were the largest seller of plant starters in New Jersey, according to their catalogs. While they didn't breed their own tomatoes, they did have their own variety of celery. Celery was a big crop in New Jersey at one time. In the 1930s, we were fifth in the nation in the production of celery. Try making cream of celery, or just about other soup without it. Or tomato juice cocktails.

There was the George Q. Hammell Seed Farms in Cedarville, which specialized in pepper and eggplant seeds. Local seeds adapted to the local growing environment.

Same thing with local seed farms, only on a larger scale. Local and regional seed saving enhances the inherent power of seeds and creates the most fail-safe approach to securing food that one can take.

Burpee's Sunnybrook Farm was an 87-acre experimental farm in Woolwich Township, NJ. It is not as well known in the Burpee story as Fordhook Farm, but its role in South Jersey seed history is significant. I've been able to examine nine of the crops lists between 1906 and 1918. A lot of trialing went on there. Asparagus roots, eggplants, okra, peppers, all types of melons, and beans were trialed. So were flowers. A lot of flowers. Flowers equal pollinators.

Of course, being the tomato guy that I am, and this book being about the Jersey tomato, my interest in examining these documents was specific to tomatoes, or in this case a specific tomato, the Burpee Sunnybrook Earliana that I highlighted earlier in this book.

From the crop lists I examined, it turns out that the Burpee Sunnybrook Earliana was the only tomato consistently trialed there. There were some trials of dwarf varieties, but their acreage was limited to one acre when grown, compared to the Sunnybrook Earliana that was grown on five to twelve acres depending on the year.

When you read how Burpee was so taken by the Sparks Earliana in his catalogs starting in 1901, it's interesting to see that he purchased a farm thirteen miles from where the original Sparks Earliana debuted on George Sparks' farm. Who's to say whether that was a motivating factor, or part of a bigger need to expand because Burpee's seed business was growing. Regardless, he claimed his place in Jersey tomato breeding history by having a farm here to grow his selection of the significant Sparks Earliana.

In 1917, 7.25% (twelve acres) of his farm was devoted to his Sunnybrook Earliana. And there's this quote from *Burpee's Annual 1916*:

> *"Choice seed of our own growing in its 'South Jersey Home' this strain is far superior to the stock usually sold, but of course, not equal to the new "Sunnybrook Earliana."*

A man of vision, William Henry Maule took over his father's lumber and seed business in 1882, and soon after split the business into separate entities. Enamored with the seed business, he kept that for himself and gave the lumber business to his brother Charles. In his 1902 catalog, Maule celebrated the 25th anniversary of his seed business, so it seems that he started with seed only in 1877.

W. Atlee Burpee had partnered with Maule's father to expand his lumber business to include seeds. Once Maule started selling only seeds, he parted ways with Burpee, but they remained friends. Burpee would eventually buy the William Henry Maule Seed Company.

In 1889, Maule changed his business model from targeting distributors to direct sales to anyone with a mailbox. He printed beautifully illustrated catalogs that included a free package of seeds and offered cash prizes for the largest orders. By 1902, he had 580,000 customers, and had distributed 3 million packages of free seeds. His silver anniversary catalog, where he took a well-deserved bow, was two years in the making.

The 1902 catalog is where the Jersey connection begins. Under a page 1 headline, *Briar Crest And Panmure*, Maule announces he's coming to the Coastal Plain soil of the Jersey peninsula terroir. See page 105 for the catalog page this is from:

> *My new trial grounds, which I shall call the Panmure Seed Gardens, are situated in New Jersey; the soil is just what I want, so light and porous that it can be worked the next day after a heavy rain. For testing all varieties of vine seeds, tomatoes and vegetables of a similar character, it is simply superb, in addition to this fact, being situated In South Jersey we can start the season fully two to three weeks earlier than Briar Crest, while in the fall the frost will not affect us as quite as early in Pennsylvania. I copy the following notice from the Newfield, N.J., Item, Nov. 22d, 1901:*

> *"An Important Sale" "Thorne D. Hallett has sold his farm, just north of Newfield, to Wm. Henry Maule Seedsman of Philadelphia. It is stated that the place will at once be out under the plow, and used as a testing ground for seeds, plants and bulbs."*

Maule's other trial ground was in Villanova, PA, about 50 miles west of Newfield.

From my research, it doesn't appear that Maule was a breeder; he secured seeds from growers and farmers around the country and introduced them under his name, or he selected for the characteristics he was looking for and stabilized the variety so the seeds would consistently grow true to those characteristics. This type of seed is known as open-pollinated.

596.

Thorne D. Hallett, et ux,

TO

William Henry Maule,

This Indenture, MADE the *fifteenth* day of *February* in the year of our Lord one thousand nine hundred and two

Between *Thorne D. Hallett and Lilian M. Hallett his wife of the Township of Franklin, in the County of Gloucester and State of New Jersey* of the first part, and *William Henry Maule, of Philadelphia, Pennsylvania* of the second part: **WITNESSETH,** That the said party of the first part, for and in consideration of the sum of *One Thousand Four Hundred Dollars* lawful money of the United States of America, well and truly paid, by the said party of the second part, to the said party of the first part, at and before the ensealing and delivery of these presents, the receipt whereof is hereby acknowledged, have granted, bargained, sold, aliened, enfeoffed, released, conveyed and, by these presents, do grant, bargain, sell, alien, enfeoff, release, convey and confirm unto the said party of the second part, *his* heirs and assigns, ALL that certain tract or piece of land and premises, hereinafter particularly described, situate, lying and being in the Township of Franklin, in the County of Gloucester, and State of New Jersey, and bounded as follows, viz.:

Beginning at a corner in the West line of the Millville and Glassboro railroad twenty five feet from the centre thereof and at the distance of twelve chains and thirty five links Northwardly by said railroad from the centre of the intersection of said railroad with the road leading from Matthews corner to the Lake; thence (1) South, sixty three degrees and thirty minutes West, twenty six chains and ten links, along the line of Alexander Sloane's land, to a corner in the line of formerly Joshua Richman's land; thence (2) by the same, South, twenty degrees and thirty minutes East, one chain and forty three links to a corner of N. H. Hallett's land; thence (3) bounding thereon, North, eighty four degrees and thirty minutes East, twenty four chains and ninety links, to a stone, corner of said N. H. Hallett's: thence (4) by the same, North, sixty three degrees and thirty minutes East, three chains and twenty three links to a corner in the West line of the Millville and Glassboro railroad aforesaid and twenty five feet from the centre thereof; thence (5) along the West line of said railroad, North, twenty six degrees and thirty minutes West, ten chains and twenty six links to the place of beginning. Containing seventeen acres of land, more or less. Being the same premises that Franklin O. Springer, Sheriff of the County of Gloucester, conveyed to Thorne O. Hallett, by deed dated March 1, 1901, and is recorded in the Clerks Office of Gloucester County in Book No. 185 of Deeds, page 146 &c.

Together with all and singular the buildings, improvements, woods, ways, rights, liberties, privileges, hereditaments and appurtenances to the same belonging or in any wise appertaining, and the reversion and reversions, remainder and remainders, rents, issues and profits thereof, and of every part and parcel thereof: And also, all the estate, right, title, interest, property, possession, claim and demand whatsoever, both in law and equity, of the said party of the first part, of, in and to the said premises, with the appurtenances. **To have and to hold** the said premises, with all and singular the appurtenances, unto the said party of the second part, *his* heirs and assigns, to the only proper use, benefit and behoof of the said party of the second part, *his* heirs and assigns, forever.

This is the deed for the 1902 purchase of the land for Maule's Pamure Seed Gardens in just north of Newfield, NJ.

The Maule Seed Business is 25 Years Old This Year.

I fully realize that this fact is of interest to many of my friends but to the public, it does not make much difference whether I have been in business 25 years or 100 years. What they are interested in is what I am going to do for 1902.

I have been working for two years on this Silver Anniversary Catalogue, with the determination of giving my friends and customers a book that would at least equal, and I hope surpass any of the twenty-four publications I have previously issued. Because the house is 25 years old, I do not want anyone to think I am getting too old to take off my coat, roll up my sleeves and push this business for all I am worth. Twenty-five years from now, if I am alive, I may think of letting up a bit; but I am only forty-four years old in 1902 and while I have worked pretty hard for 25 years past, I still propose to keep the Maule Seed business right in front of the procession. It is fortunate, not only for myself, but for my customers as well, that I made such elaborate preparations for my Silver Anniversary Book. Notwithstanding short crops, I go into the season with by all odds the largest stock of the very best seeds I have ever warehoused, and on account of these large stocks, I am prepared to quote Maule's Seeds at most reasonable prices, considering the remarkable shortage of a great many seed crops, both at home and abroad.

I do not propose to say much about the past 25 years; the illustration on page 32 taken over 20 years ago, and the illustrations on this page and page 2 of my present warehouses, will give everyone some conception of the astonishing growth of this business; but to give some further idea of what I am doing, I would say, that last year's postage bill paid Uncle Sam amounted to $37,273.19; and in addition I shipped double the quantity of prepaid packages by express of any other house in Philadelphia, and more than any other seed firm in the country. During the last 25 years, I have paid in cash prizes for club orders and premium vegetables $29,909.98. I have distributed during the last 25 years more than three million packets of Maule's Seeds free for trial among my customers.

During the existence of this business more than five million seed catalogues have been distributed, in addition to many million pamphlets, circulars, and other printed matter. Since 1889 Maule's Seeds have not been sold to dealers; but can be obtained only direct from headquarters in Philadelphia. I am not only the original American house adopting this method; but for many years no other house dared follow my footsteps.

All these facts are well known to my old friends, and they are simply reiterated here for the benefit of new readers; but I cannot forget under any circumstances to thank the more than 560,000 customers whose names are now on my books, for the share each individual one has had in building up this enormous business. The kind words spoken by my friends to their friends and neighbors the last twenty-five years, have been the best advertisement this business has ever had, in fact have been the foundation of the whole structure, and in this, my Silver Anniversary Year, my one regret is that I cannot take each one by the hand and thank them personally.

For 1902 I hand you a book I am proud of. Never in the history of the seed trade has a single house been able to present anything like the aggregation of new things I have to offer my customers this year. These yellow pages are simply filled to overflowing with the choicest lot of the most desirable novelties in vegetables ever gotten together in the space of a single book; while the other departments, flower seeds, plants, bulbs, etc., also contain everything new or old known to the trade worth growing. For my Silver Anniversary Year I have made preparations for the largest business I have ever done, and trust I will not be disappointed. Wishing you all a prosperous New Year, I remain   Yours to command.

Wm H Hny Maule

BRIAR CREST AND PANMURE

My trial grounds at Briar Crest have had a national reputation for years. They are unquestionably the most thorough in America; every thing is tested and tested well. I have frequently discovered things there of vital importance that have been overlooked by every other trial ground and experimental station in the country. It is by reason of these trial grounds and thorough tests, that I am able to give my customers each year the best of everything worth growing and my friends have made hundreds of thousands of dollars by following my advice and planting largely of new varieties I have introduced. More actual cash has been realized by my customers with new varieties of my introduction than by the customers of any other house.

The one fault about Briar Crest has been the soil. It is the heavy Pennsylvania soil of Montgomery Co., that has produced forty years ago as high as 50 bushels of wheat to the acre, and while it is the very best for the purpose for many things, yet being of this heavy character there are some things that can be tested better on a lighter soil. With this end in view I have been looking for some time for a piece of land that would answer my purpose in this respect, and am glad to report that I have at last been able to secure it.

My new trial grounds, which I shall call the Panmure Seed Gardens, are situated in New Jersey; the soil is just what I want, so light and porous that it can be worked the next day after a heavy rain. For testing all varieties of vine seeds, tomatoes and vegetables of a similar character, it is simply superb, in addition to this fact, being situated in South Jersey, we can start the season fully two to three weeks earlier than at Briar Crest, while in the fall, the frost will not affect us quite as early in the season as in Pennsylvania. I copy the following notice from the Newfield, N. J., Item, Nov. 22d, 1901:

"

"Thorne D. Hallett has sold his farm, just north of Newfield, to Wm. Henry Maule, Seedsman of Philadelphia. It is stated that the place will at once be put under the plow, and used as a testing ground for seeds, plants and bulbs. Mr. Maule has trial grounds near Villa Nova, Pa., where thousands of tests are made every year. Vegetables and flowers, especially new varieties, are thus tested and proved before they are catalogued for sale. In addition to the novelties produced in America, each season, many new things come from abroad, especially from Germany, France, England and Japan. Leading seedsmen are always on the lookout for vegetables and floral treasures, and depend upon their trial grounds to determine them. Mr. Maule comes to New Jersey to take advantage of our warm, light soil. He will retain his Pennsylvania trial grounds, and thus have the combined advantages of two testing stations."

Briar Crest and Panmure Seed Gardens will be run in connection, one with the other, and on account of the variation of soil and climate, I can safely say there are no trial grounds in America better situated for testing purposes. With Briar Crest alone, I considered there were very few discoveries made in other trial grounds or experiment stations that were not noticed at Briar Crest; but with both Briar Crest and Panmure I consider I lead them all.

MARKET GARDENERS. My Special Wholesale Price List for Truckers will be mailed on application; but it must be distinctly understood that this Price List will not be sent to, nor will any orders be filled from it for anyone not in the business of raising truck for sale.

A WORD TO NEW READERS OF THIS BOOK.
Is there a gardener in your neighborhood who is always first in market, and always captures the premiums at your Fair? If so, ask him about Maule's Seeds; the chances are he knows all about them.

MAULE BUILDING, MARKET AND EIGHTEENTH STREETS

(COPYRIGHTED. ALL RIGHTS RESERVED.)

This is the catalog page with the announcement of the purchase of the land for Maule's Pamure Seed Gardens in just north of Newfield, NJ.

Since we're talking tomatoes here, let's look at Maule's Success tomato. We'll start in the 1900 catalog and his *"Maule's Novelties and Specialties for 1900"* section. He introduces *"A BRAND NEW TOMATO. Maule's '1900.'"* in grand style. He devoted a full page in this section to this new tomato, complete with a large and prominent illustration that identifies the source of the seeds in the caption as Mr. Miesse, who is holding a long vine loaded with his famous tomato. Mr. Miesse was M. M. Miesse, a market gardener who was a tomato specialist from Lancaster, Ohio. Miesse is credited with the origination of this 1897 variety that Maule commercially introduced in 1900.

Maule liked contests, so there was one with $600 in prizes offered. There were six different contests, each with a $100 prize. The contests included the largest specimen, the finest specimen, best testimonial written on a postcard, best name (since Maule's 1900 was just a number to differentiate it from other tomatoes), and the largest order of $.50 orders. Each winner would receive a free package of 1900 seeds, but each order had to go to a different address. He also offered a free package of the 1900 to any order over $.50, one per customer. If more than one package was desired, then they could be bought at a discount.

He distributed 70,000 seed packages of Maule's 1900 tomato in 1900 and received almost 10,000 letters praising this tomato. Seventy thousand seed packages; that's a lot of seeds. He published almost a full page of testimonials about his Maule's Success tomato, and 37 people sent in Success as the name. In 1901, Maule announced the new name for the 1900 as Success.

Maule wasn't the only seedsman offering Success in 1901. Burpee offered it in 1901 also. Without mentioning Maule by name, he wrote about the Nineteen Hundred tomato, dropped Maule's name from the tomato, and just used Success when referencing 4,000 testimonials from forty states and territories that Maule would publish in his 1901 catalogs. Burpee did recognize Miesse as the originator, however, he got the first name wrong. He cited Wm. Miesse when it was M. M. Miesse. While Burpee said it had 4,000 testimonials, Maule states in his 1901 catalog that he received about 6,000 letters. Atlee Burpee and Wm. Henry Maule were friends, so that's likely how that information found its way into Burpee's catalogs. In my research at Smithsonian Gardens, I've seen personal papers noting the relationship. The end of the Wm. Henry Maule Co. occurred in 1947, when Burpee acquired the company.

I grew the tomato for the first time in 2024, and I can see why it's a good market tomato. The plants are strong, prolific, and produce a flavorful tomato. It doesn't have the acidic note that Jersey tomatoes have, since it originated in Ohio.

A BRAND NEW TOMATO. MAULE'S "1900."

$600 IN CASH PRIZES.

A PACKET FREE OF THIS COMING LEADER to Every Purchaser of 50 Cents Worth of Goods from this Ca[t]alogue, Six of Whom Will Each Receive $100, Nov. 1, 190[0].

At an enormous price, which I venture to say has seldom, if ever, been paid for a new vegetable, I have secured from Mr. Miesse the privilege of introducing to the world at large his new tomato, which for all purposes will prove the coming leader either for forcing under glass or culture in the open ground, and is bound to make a name for itself among tomato growers everywhere. Besides being the best tomato for forcing purposes I have ever seen, after Enormous, its super[i]or for a late crop does not exist. It is exceptionally fine for cann[ing] purposes, being always of a beautiful, bright, brilliant red color. [I] have never put out any variety with greater confidence, and on [ac]count of the superior merits of Maule's "1900" have decided not o[nly] to give the variety a full page in my annual catalogue, but also to o[ffer] the unheard of amount of $600 in cash prizes for a single vegetable [in] one year, under the following conditions:

$100 for the largest specimen of fr[uit] without regard to shape.

$100 for the finest specimen of fr[uit] without regard to size.

$100 for the best testimonial and [de]scription to be written on a p[os]tal card, for publication in [my] 1901 catalogue.

$100 for the best report of a comp[ar]ative test between "1900" and [the] greatest number of other va[rie]ties of tomatoes.

$100 to the customer who sugg[ests] the best name for this blush[ing] beauty, as "Maule's 1900" is [not] its name, but simply a num[ber] by which to designate it fr[om] other varieties.

$100 to the person sending me [the] greatest number of 50-cent ord[ers] from this catalogue, each [of] which will secure a packet [of] Maule's "1900" free, and be [en]titled to compete for the ab[ove] prizes. *Each order to go to a [sep]arate address.*

A PORTRAIT OF MR. MIESSE AND HIS FAMOUS NEW TOMATO "MAULE'S 1900."

Terms of Competition.

These prizes can be competed [for] only by those whose names are re[gis]tered on my books as customers, [and] have either received a packet [of] Maule's "1900" free with a 50-cent or[der] or purchased a packet of the seed.

All reports and specimens of f[ruit] sent in competition can be forwar[ded] at any time up to Nov. 1, 1900, when [the] prizes will be promptly awarded [and] paid in accordance with my usual [cus]tom. Nothing sent after this d[ate,] however, will enter the competitio[n.]

The plants grow between a standard and a dwarf with very short joints, and a large cluster at every second joint; the tomatoes hanging one cluster on top of another right up the plant, so much so that the tomatoes themselves completely conceal the stems; these clusters contain 6 to 10 tomatoes and all ripen all over and through at the same time. Like any unusually solid variety it contains very few seeds, much less than the Stone and other standard sorts; of course, every one will recognize this is a very strong point in its favor. Its habit of growth is excellently shown in the accompanying illustration taken from a photograph, which not only shows the single tomatoes, clusters, etc.; but is also an excellent portrait of Mr. Miesse himself, who is now celebrated from one end of the country to the other as the originator of the Enormous Tomato, Emerald Cucumber, White Cob Evergreen Corn, and half a dozen other popular and profitable varieties. Mr. Miesse has for a number of years fruited each winter 2500 plants under glass, sold the product at fancy prices, and has averaged a net price of $250 per season from each 100 foot of house. Mr. Wm. J. Green, bor[t]iculturist of the Ohio Experimental Station. Wooster, O., after a v[isit] last year to Mr. Miesse's place, wrote me as follows:

"I was very much impressed with the vigor and fruitfulness of [this] variety, and believe it will be most satisfactory for forcing purpose[s.]"

Last summer from the late crop in the open ground. Mr. Mi[esse] supplied all the hotels and leading families in Lancaster, O., with f[ruit] of this variety for several weeks. The general opinion of all his cu[sto]mers, as well as of all those who tested this variety at Briar Crest [the] past summer was unanimous, that Maule's "1900" was absolutely [the] finest flavored tomato they had ever eaten, either raw or cooked. [To] give an illustration of its superior quality would say, Mr. Miesse w[rote] me last September that the proprietors of the various hotels [in] Lancaster, O., to whom he furnished these tomatoes, reported t[hat] all commercial men stopping with them seemed to be unusu[ally] impressed with the superior quality of his new tomato, and invaria[bly] commented upon it, although naturally their knowledge of tomat[oes] was limited.

Now for My Astonishing Offer.

Notwithstanding the fact that this New Tomato is the most startling novelty of the year, I will present **A PACKET OF SEED [OF] MAULE'S "1900" TOMATO FREE** to every customer ordering goods from this catalogue to the amount of 50 cents or over; **but no m[ore] than one packet will be presented to a single customer, no matter how many orders they send.** If, however, any of my frie[nds] would like more than one packet, the price will be

25 cents a single packet; 5 packets, $1.00; 12 packets, $2.00.

One final note about Maule, from my perspective as an award-winning graphic print designer: Maule's catalogs were exceptional in their concept and use of design. The pages used superb illustrations that showcased the texture of the subject more than other catalogs I've looked at. And he used a lot of illustrations per page, and did so with the right proportions so they didn't overwhelm the pages. A lot of pages are packed with type, but working within the constraints of typesetting at the time, they were successful.

He illustrated each floor of his properties in Philadelphia. One illustration, titled *Sectional View of his Main Establishment, 1711 Filbert Street*, shows each floor with captions indicating their function. His "*Section View of Maul Building, Market & 18th Sts, Philadelphia*" shows each floor without captions, but they're not needed since the illustrations tell the story so well. The catalogs with illustrations feel more like storybooks to me than catalogs. And my favorite seed catalog cover is his *1896 Seed Catalog*. It uses an artist's palette to display some of the sweet peas that he was offering that year. For me, Mother Nature is the ultimate artist, and this illustration demonstrates that like no other seed catalog illustration I've seen.

No history of Jersey tomato seed breeding would be complete without the Johnson & Stokes Seed Company. Launched in 1878 by Herbert W. Johnson, who operated the Johnson Seed Co., in Moorestown, NJ, it became the Johnson & Stokes Company in 1881 when Walter P. Stokes joined as a partner. Johnson left the partnership in 1906. In 1907, the first Stokes Seed Farms Catalog was published.

Francis C. Stokes, Walter's son, took over in 1916, and the name was changed to Francis C. Stokes & Co. Francis was a Rutgers Horticultural alumnus who developed the following Jersey tomatoes: the Stokesdale, Stockdale #4, and improved selections of Bonny Best and Geneva John Baer. He innovated outside of the greenhouse as well by being the first seedsman to offer seeds in a tin can and the first to protect seeds with a fungicide. ·

Johnson & Stokes Seed Company introduced the Atlantic Prize tomato in 1889, Sparks Earliana in 1900, which became a mainstay of the tomato canning industry, and Stokes Bonny Best in 1906, so when Francis took over in 1916 he was building on an established legacy. Francis C. Stokes & Co. went on to introduce their selection of Marglobe, Master Marglobe, in 1930, and their Stokesdale in 1936.

As a side note, Stokes commercially developed the seed release of the USDA-bred Marglobe in 1926.

For the 1936 growing season, Francis C. Stokes & Co. made a big decision to focus only on tomatoes. From the 1936 *Stokes for Tomatoes* catalog:

For the last ten of our fifty-four years as seedsmen, our principal interest has been in Tomato breeding. We have now decided, obviously with mingled feelings, to devote our attention completely to Tomato products, leaving the enormously complex problems attendant on the breeding of other vegetables to those who by training, technique, and location are better suited to handle them. (This decision does not immediately affect our southern branches.)

We are quite content to let others take the strings out of beans, the cores out of carrots, and the curves out of cucumbers. We will devote our energies to the development of bigger and better, deeper and sweeter Tomatoes. That in itself, we think, is a man-size job. In this connection we would remind you that it is much less than 100 years ago that the despised Love Apple was looked upon as deadly poison. Today it is the Number One vegetable in the United States.

With this announcement of a pivot to tomatoes only in 1936, they introduced their Stokesdale. The Stokesdale was found in a section of Bonny Best in 1933. They didn't know if it was a genetic mutation or an accidental cross between Bonny Best and Marglobe. They just knew they had an early tomato of exceptional quality. The Stokesdale would become the focus of their drive to breed the best Jersey tomato in the state.

They went all-in with hybridization of tomatoes with their Stokescross line, and did two breedings a year. Stokescross Hybrids represent a big step forward for the tomato industry.

While reading their 1962 *Stokes' Tomato Seed Red List Prices For Seedsmen and Dealers,* I came across the following under a subheading:

Our seed business was restricted to tomatoes only. Ours is one of the very few firms in this country that has specialized in a single vegetable. In 1898 the old firm listed 38 varieties of tomato. In 1962 we carry 10 varieties. These are carefully described in this folder. At the turn of the century, 5 tons per acre were considered a full crop. In 1961 our growers produced over 20 tons per acre.

We are, of course, aware that our 10 varieties do not cover the entire field, but as Tomato Specialists, we prefer to confine our breeding and growing projects. You are invited to inspect our plant at Vincentown, 20 miles from Philadelphia, and see the growing fields that surround us.

I checked their 1898 catalog and indeed there are 38 tomatoes listed, including a contest for naming a new tomato. In 1899, there were eleven winners of the name-the-tomato contest. The winning name was The Early Bird.

Their list of tomato seeds on offer in a 1962 seed list pamphlet included two Campbell's Soup varieties, including the Campbell's 146 highlighted in this book. It was a rare public offering of Campbell's-bred tomato seeds, since Campbell's didn't release seeds to the public.

Two tomatoes from Willard Bronson Kille, the noted tomato breeder from Gloucester County, NJ, are listed: Kille #7 already highlighted in this book, and Killie #18 that, according to this pamphlet, was the *"finest of his selections."*

Stokes through the years would offer selections from local farmers, and in this case, they offered a sample from Campbell's historic breeding program. They offered their legacy varieties and seed from a noted local tomato breeder with his own tomato experimental farm. Quite a legacy.

Once I found the depth of the historical connection of canners, seeds, and farms, I learned the meaning of gobsmacked. I found myself looking at a personal disconnect about the power of seeds even after working with them for years.

When I started working with them, I didn't fully understand or appreciate them. I certainly didn't appreciate their profound existence. I threw them away when preparing food. I ate roasted pumpkin seeds by the handfuls with my aunts and cousins at my grandmother's kitchen table. When shaken out of a bulb-shaped glass container at a pizzeria or Italian restaurant, seeds were about spicing my food. I never gave a thought about all the seeds involved with my food.

While researching this book, I found an interesting business model with seeds and canning being part of the same company. It's smart: a dual revenue stream from one crop. From a seed and food sovereignty point of view, this is a strong relationship. It's a shame that they got separated, but large food companies made the decision to sell off the seed assets when they were scooping up the smaller local processors. This happened a long time before seeds became commodities and intellectual property. Thanks to a bad 1980 U. S. Supreme Court decision,

Diamond v. Chakrabarty, about patents and living organisms, an avenue opened for the intellectual property claim of seeds and genetically modified organisms (GMOs) in the 1990s.

The seed and canning history in South Jersey shows how this combination was an economic benefit to local communities, while providing seed and food sovereignty. Folks then didn't realize what they had with seed and food sovereignty, since it was all local. It was everyday life.

Campbell's Soup is an interesting case. I want to be very clear here: I'm looking at their tomato breeding program from a cultural seed and food sovereignty point of view, and that's it. It's not a defense or support of their business choices.

They are a major food manufacturing business built in large part on their tomato seed breeding. They bet big and early on agricultural research, which helped make them a global corporation. Yes, they were pushing the newest convenience foods, while they were practicing seed and food sovereignty in that they were creating culturally appropriate seeds and foods. They contributed significantly to the development of the Jersey tomato and the local culture of it.

They were doing it without corporate interference, because they were the company creating the seeds. They had a high school farmers' recruitment program. They did field trials with small family farms that grew for them also. The farmers got paid. Generations of families worked for Campbell's. They were a union shop when their food production was here. Local food was secured through their canning process. They gave plants to their employees, who could eat, share, and can their harvest from their yards.

They accelerated the local innovations in tomato breeding, canning, and glass. Across Jersey, these innovations were happening at smaller companies, too. Jersey tomato culture was a way of life. A folklife and foodway. It was reflected in the quantity of tomato products that were the pride of local farms and seed and canning companies.

Campbell's Soup didn't own farms and they had a closed system of seed distribution, so the sovereignty of their seeds was private. That is a major problem for the sovereignty of the seeds they bred. From what I have been able to determine, they didn't restrict seed saving. I have Campbell's 146 seeds that were saved by generations of a family whose relatives worked at Campbell's. They saved seeds in case Campbell's stopped distributing seeds. Also, I found Garden State tomato seeds, a Campbell's variety, in a local seed catalog from 1955. And from 1960s seed lists, the Stokes Seed Company was selling seeds for a couple of varieties of Campbell's tomatoes.

They are undeniably a part of Jersey tomato culture. Innovation was their niche with condensed soup. That led to tomato breeding.

John Thompson Dorrance was hired by the Campbell's Soup Company in 1897. Dorrance was the nephew of Arthur Dorrance, Campbell's General Manager. This hire would become a pivotal event in the history of food processing and Jersey tomato breeding. Dorrance invented the condensed soup process. He also had a farm—and a tomato named after him, the J.T.D.

This is what the *United States Department of Agriculture Yearbook of Agriculture, 1937* had to say about the J.T.D tomato:

> *The J. T. D. is an interesting example of a local type developed for adaptation to a specific set of conditions and needs. It was developed by the Campbell Soup Co. for growing in New Jersey, mainly for its own factory use. It has not become widely grown elsewhere.*

In 1928, the J.T.D. tomato was used in a breeding program for what is the most popular Jersey heirloom tomato, the Rutgers. The J.T.D. was crossed with the Marglobe, a historic tomato developed by Frederick John Pritchard while at the Plant Industry Bureau at the USDA.

The Marglobe has strong disease resistance to nailhead rust and Fusarium wilt. Nailhead rust was a big problem in the Florida tomato industry. Because Marvel, one the parents of Marglobe, was resistant to it, that resistance was bred for. As a result, Marglobe became a parent for many varieties and was likely the reason Campbell's used it for the Rutgers. While the Marglobe isn't Jersey-bred, the seeds were commercially developed here. It's a parent to the beloved Jersey heirloom Rutgers, so to me, it's an honorary Jersey tomato. The Garden State Improved variety is from crossing the Garden State with the Marglobe.

The Rutgers tomato was developed by Professor L. G. Schermerhorn of Rutgers University's New Jersey Agricultural Experiment Station, and Campbell's. It was used by Hunt's and Heinz in their tomato products. Thomas J. Orton, a professor in the department of plant biology and pathology at Rutgers University, claimed it was the most popular tomato in the world before mechanical harvesting of commercial crops started in the 1960s. While it lost favor in commercial fields, it never left the backyard gardens throughout the state. It's a great tomato.

Campbell's kept with their tradition of New Jersey beefsteak tomatoes. They bred one in the Rutgers. Before they introduced their condensed soup process, they canned their Beefsteak Tomato Soup. They also bottled a beefsteak ketchup.

Innovation in South Jersey in the late 1890s and early 1900s wasn't limited to condensed soup and Jersey tomato breeding. Edgar F. Hurff, seedsman, farmer, and canner, created his own path in the loamy soil in Woolwich Township and Swedesboro, NJ.

Hurff started out as a tenant farmer. His landlord suggested that he start his watermelons a month later than everyone else so Hurff would have melons when no one else did. That plan got interrupted by a storm that brought in cold weather. Hurff couldn't sell his melons for good prices. He left them in the field.

The next year, Hurff did plant his watermelons late. In October of that year, the president of the D. Landreth Seed Company, in Philadelphia, PA at the time, was out and about. He saw Hurff's field of melons, and asked what he was going to do with them. Hurff replied nothing and agreed to sell the seeds to Landreth.

Unfortunately, D. Landreth went into receivership before Hurff could process his seeds. Hurff reached an agreement with Landreth: if he would give them seeds, he would be paid the following year. When Hurff went to collect payment the following year for the melon seeds, he got a contract for tomato seeds to be saved in 1902. That began a long relationship with D. Landreth until Hurff sold his business in 1945. During the course of this business relationship, Hurff sold watermelon, cantaloupe, cucumber, pumpkin, squash, eggplant, pepper, and tomato seeds to D. Landreth.

Hurff kept hogs on his farm and fed them the waste from seed processing. As his seed business grew, the waste increased, and so did the need for more hogs. In the Jersey spirit of innovation, he started a canning company to process the leftover waste from seeds. In 1913, he opened his canning factory. Addressing food waste in the early 1900s: way to go, Edgar.

In his first year, Hurff canned tomato products, ketchup, and pickled peppers. In 1915, he sold his 200-acre farm so he could concentrate on his canning and seed business. He canned tomato, vegetable, pea, and bean soups; mixed products like spaghetti and egg noodles with tomato sauce and pork and beans; and squash and pumpkins, potatoes, kidney beans, and pears. He sold processed tomatoes to Heinz and Campbell's Soup. He even canned dog food.

Hurff advertised nationally. The ad on the right is the inside cover of *Life* magazine from November 1, 1937. This was a premium media buy. As you can see, the tagline on his company's logo reads "*The Jersey Tomato People.*" Part of the copy reads:

> "*… But when it comes to garden truck (especially tomatoes) we Jersey folks claim that we're 'tops'— and so we specialize on the things in which we're specialists.*"

While not a seed breeder, I did find a series of seed lists from his business in my research. In 1930, he offered Hurff's Special Large Red Superior for canning. From what I've seen so far, it was the only year he offered it. I've read seed lists from 1928–1933 and a couple of years before and after, it wasn't offered. At the height of the seed business, he served 2,700 seed customers.

Hurff sold his business to his son and a non-family business partner in 1941. In 1958, the canning business was sold to Del Monte. They didn't want the seed business; as a national company focused on canning only, they had no need for it. There was social change afoot, kicking the seed and canning company concept aside.

Not to be overshadowed in Jersey tomato seed and canning history, Philip J. Ritter (P.J. Ritter) came to Bridgeton, New Jersey by way of Kensington, Pennsylvania. Kensington is now a neighborhood in Philadelphia. Ritter migrated to Kensington from Germany, where he started a confectionery business. His wife Louisa demonstrated her prowess with canning with her Mrs. Ritter Plum Preserves. Soon the business grew to more than confectioneries, and other fruit preserves were added, all sold under the name P.J. Ritter Conserve Co.

Because he couldn't find one to his liking, ketchup became a focus of P.J. Ritter's expansion from fruit products. He brought in a ketchup expert from Germany who developed a recipe to his liking. Ritter at this time had five kitchens servicing his Ritter label, and a decision was made to consolidate them all into one in Bridgeton.

Bridgeton offered water transportation to receive tomatoes for his ketchup, and railroad access for outbound shipping. It also offered farmland where the Ritter company set up their experimental farm to start breeding their own tomatoes and peppers. That breeding carried forward what the P.J. Ritter company decided early on: that their company would be a packer of specialty foods. Having proprietary vegetables for your own products ensures a unique taste.

P.J. Ritter used a lot of California fruits for his preserve label, and in 1891, he went into partnership with Richard Hickmott of Oakland. In 1892, they packed their

Down here in the country a soup has to be good enough to be a meal in itself!

Soup is a natural farm dish. In the country is made from good, rich soup stock and fresh ome-grown vegetables. It isn't just a fancy com- ination of rare tidbits. And—it must be a lot more ian merely "a tasty appetizer".

ountry soup was made famous by women who ok plenty of pride in their cooking and had enty of garden-fresh ingredients at their disposal. hese women faced the problem of feeding large milies of hardworking grown-ups and active, owing kids, with tasty, substantial satisfying od that was inexpensive. They solved it by atching an unlimited knowledge of soup making ;ainst decidedly limited budgets.

ow we farm folks in South Jersey are particularly essed with soup-producing vegetables and our omen folk are especially proud of their tomato, getable, pea and bean soups. That's why we ecialize on canning those four kinds.

e think, for instance, that folks in Baltimore iow more about oysters than we do. And we're willing to admit that people out in Idaho have us licked on the raising and the cooking of potatoes. But when it comes to garden truck (especially tomatoes) we Jersey folks claim that we're "tops"—and so *we specialize on the things in which we're specialists.*

We know that other soup makers have w: tempting advertisements and have put out swell soups. But all we say is: "Try Hur vegetable soup, pea soup, bean soup and to soup!" We make only these four kinds—b. make each of them good enough to be a meal in

Hurff soups are ready-to-serve just as they from the soup kettle. Folks call them *two-way* because they can be served two ways. As a w some *meal* our customers use them just as come from the cans. For a soup *course* they : half can of water to each can so that a s twenty-two ounce can of Hurff soup serves six to eight people. Richer in flavor and bodied, the addition of water does not ir their taste. And—if that doesn't mean econ we don't know what does!

HURFF *Ready-to-Serve*

SERVES 6 to 8

When Hurff Ready-to-Serve Soup is served as a course most women add ½ can of water to each can. A 22-ounce can can be made into 33 ounces of soup.

Most brands of condensed soup according to directions can be diluted by adding a full can of water. This 10½-ounce can makes only 21 ounces of soup which serves only 4 to 5 people.

Other ready-to-serve soups serve only 4. They have not sufficient richness and body to permit the addition of water. A 16-ounce can makes only 16 ounces of soup.

Ask for HU

Tomato Juice . . .
Juice Cocktail . . .
Catsup...Tomato S
Tomato Puree . .
Aspic . . . Pork and
. . . Vegetable So
Spaghetti . . . Squa:
Bean Soup . . . Pea S
Pumpkin . . . Hominy

:ASONED, COOKED AND PACKED BY EDGAR F. HURFF COMPANY

HURFF

The Jersey Tomato People

SWEDESBORO,

first canned asparagus. This led to P.J. Ritter becoming at one point the largest packer of asparagus in the East.

P.J. Ritter packed their asparagus in glass. That set them apart from the tins of asparagus on the shelves in markets. Asparagus was second in volume to the tomato products at P.J. Ritter.

The efficiency of the Ritter packing process brought the price of it down, making it accessible to the general public.

This efficiency was also demonstrated when the P.J. Ritter Company became the first processor to put ketchup on an assembly line in 1914. Since they brought over a German ketchup maker to refine their recipe, and they were first in packing ketchup on an assembly line, tomatoes for their ketchup became a concern.

William H. Ritter, the son of P.J. Ritter, who followed his father into the canning business, went to the D. Landreth Seed Company for help. William also worked with the New Jersey Agricultural Experimental Station on a five-year tomato improvement project that ended in 1938. It is interesting to note that the development of the Rutgers tomato was happening at the same time.

While the work with the Agricultural Experimental Station ended, the breeding and improvement went on at the Ritter Experimental Farm in Bridgeton. Their seed business grew substantially, and eventually was spun off into The Ritter Seed Company, which became the largest supplier of certified tomato seed in the world in the 1950s.

They also bred peppers that they used in their relishes, including their Chili Sauce Relish that was developed from one of Grandma Ritter's recipes. It was a green tomato relish and used no cucumbers. It was a smart use of potential food waste— leftover green tomatoes at the end of the season— to make a product. Their custom-bred peppers, which no other food company was using at the time, matched the Jersey cultural spirit of bold and authentic action.

Innovation in vegetable breeding and efficient glass packing, along with strategic partnerships, contributed to P.J. Ritter's growth and success. P.J. Ritter contracted with the McIlhenny family of the Tabasco brand, for exclusive right to use Tabasco product when P.J. Ritter introduced DeLuxe Ketchup. It was a specialty version of their ketchup line seasoned with Tabasco. At one time, they were the sole distributor of bulk Tabasco products in the U.S.

P.J. Ritter was bought by Curtice-Burns in 1967. Curtice-Burns was the result of consolidation of a small and regional producers in New York who were looking at ways to maximize their older and smaller operations, since California was becoming the new hub for tomato products. Curtice-Burns was a cooperative owned and run by the farmers. They also purchased Brooks Foods in Collinsville, Illinois. Curtice-Burns made ketchup and when they purchased P.J. Ritter and Brooks they discontinued their own product. Ritter continued to produce ketchup under the Brooks Foods label until they closed in the mid '70s.

Ketchup, and Jersey Canning History

The more I researched the canning industry here, the more I learned about the prominent role of the Jersey tomato in the rich and vibrant history of ketchup. You already know that the P.J. Ritter Company of Bridgeton, NJ, was the first processor to put ketchup on an assembly line in 1914. Heinz and Salem County, NJ, have a long history not only with ketchup, but with all tomato products in their 57 varieties. Salem County also had many small canning operations. Remember that the Sparks Earliana tomato debuted in the heart and soul of Salem County soil, so the Jersey tomato has deep roots of love and passion within Salem County.

I fell into a deep and wonderful rabbit hole researching ketchup. As I started down the ketchup research path, I thought I would have an ancillary experience with ketchup history: some key facts and I would be done. That, however wasn't the case. I found a fascinating and rich history that, to me, has an edge to it. One of the many reasons I love research.

The descent down this rabbit hole started with the rapid commercialization of tomato ketchup. Canned products were needed for soldiers during the Civil War. Once the war ended, the canning industry continued to ramp up. Ketchup was one of the tomato products that these companies produced. Jersey, and the Jersey tomato, played a large role in that.

A little background on ketchup. It's not exactly a linear history. There are flavoring sauces that use preservation and fermentation processes that go back to ancient Rome. As England started colonizing, it also colonized recipes and processes. There's documentation of ketchups being brought to England before 1680. These were sauces, not the condiment we use today. They started to gain popularity in England in the 1700s. England developed its recipes using walnuts, mushrooms, and fish. These recipes followed the British when they colonized America.

Tomato-based ketchup appears in the ketchup history timeline in the early 1800s, and the first known tomato ketchup recipe was published by the American scientist and horticulturist James Mease of Philadelphia in 1812.

As tomato canning and the production of ketchup expanded rapidly after the Civil War, Camden and Philadelphia, along with Cincinnati and St. Louis, were the southern boundary of the core ketchup-producing area. New York City and Rochester, were the northern boundary, with Detroit, and Chicago as the boundary to the west. The following is an overview of New Jersey ketchup production during the rapid expansion. It's by no means complete.

The first recorded production of ketchup in New Jersey happened at the North American Phalanx, a secular, socialist Utopian society in Red Bank, NJ. Despite its noble intent, the society didn't last very long. There was a farm and a canning operation. In 1855, John Bucklin produced a large quantity of ketchup there.

In 1862, Abraham Anderson started a canning company, which would become Campbell's. By 1874, Joseph Campbell joined him as a partner, and they trademarked the word "Beefsteak" tomatoes with an illustration of two men holding one large tomato. Anderson and Campbell eventually parted ways. As part of their new partnership, when Arthur Dorrance joined Campbell's, they retained "Beefsteak Tomato Ketchup."

While Campbell's relied on condensed soup, they continued to private label manufacture ketchup for a couple of companies in Philadelphia, and one in Alabama. They continued to make ketchup into the twentieth century, but it wasn't the main focus of the company.

Edgar Hurff made ketchup that he advertised as made with pineapple vinegar.

In Woodstown, NJ, B. S. Ayers & Sons had a ketchup plant. They claimed that in 1909, they bottled one million bottles. I live in Woodstown and there's not a trace of this history here. In 1913, Edward Pritchard, who was bottling ketchup in New York City, bought Ayers & Sons and moved it to Bridgeton, NJ.

When Pritchard arrived in Bridgeton, the P.J. Ritter company was already there making ketchup. As noted earlier, the P.J. Ritter Company saw this potential and brought in a ketchup expert for a recipe. They were the first to put ketchup on an assembly line in 1914 and breed tomatoes to achieve a higher quality raw material.

Heinz set up shop in Salem, NJ, in 1906, processing tomato purée. They started making ketchup in 1909. They kept their plant there until the mid 1980s.

And finally for this brief recap, there was the Moorestown Canning Factory in Moorestown, NJ. They opened shop in 1868, producing thirty thousand gallons of ketchup annually. Eventually, they were bought out by one of the grocery stores that sold their product, who continued to have ketchup made in Moorestown.

There was a lot of canning going on in New Jersey. The canning industry in New Jersey was diverse. While tomatoes and tomato products were a big part of it, canning here included pumpkin, squash, Kieffer pears, pickles, caviar, cherries, berries, beans, beets, and rhubarb. All locally grown, and enough variety to stock your pantry for the winter. Food sovereignty at its best.

To show the density of canning companies, I'm going to start with a list published by the U.S. Department of Agriculture in 1921, based on a soil survey from 1917. Millville is a small city in Cumberland County, NJ. It's a hub for the Bayshore area of Jersey, the southern coastal area along the Delaware Bay and the inland coast of the Jersey Cape. I found this document while researching at the Stockton University archives, titled, *Soil Survey of the Millville Area, New Jersey.*

A sample of the depth of the canning industry in South Jersey in 1921	
Bridgeton There were fourteen plants. Nine canned all tomatoes, the other five processed Kieffer pears, pumpkins, beans, squash, beets, and rhubarb.	**Goshen** There were two canning plants processing tomatoes, pumpkins, and Lima beans.
Cape May There were six plants, mainly processing tomatoes, Lima beans, and peas.	**Landisville** There were two canning plants. Both canned tomatoes, and one also processed small fruits and peaches.
Cedarville There were two, one that canned peas and tomatoes and, the other tomatoes, peas, and beets.	**Newport** There was one tomato canning plant.
Egg Harbor There was one canning plant, tomatoes and pears	**Rio Grande** There was one tomato canning plant.
Eldora There were two canning plants tomatoes, pumpkins, and Lima beans.	**Vineland** There was one plant canning tomatoes, peaches, and berries.
Fairton There was one plant canning tomatoes and cherries.	

The area for this soil survey includes parts of Cumberland, Salem, and Atlantic counties, and all of Cape May County. It's 1,002 square miles, which sounds like a lot, but if you were in the center of the Bayshore area, say in Bridgeton or Millville, you wouldn't be more than an hour away from any given area. That hour is because of how far it is to two towns at the southern part of the Jersey cape.

I started reading the soil survey since soil and seeds go hand in hand. Soil surveys have agriculture information because, again, soil and seeds. On page 10, the last paragraph in the "Summary of the Area" section detailed the information about the canneries in the Millville area. To my surprise and delight, it also listed the products canned, shown in the previous table.

Thirty-three canneries. Keep in mind, this is not even for all of South Jersey. And a majority of this area is outside the Coastal Plain Belt of soil, which is the area with some of the best soil in the country for growing crops.

New Jersey's first cannery was in Burlington County, and it canned herring roe. One of the first canneries to successfully can tomatoes for commercial use was in Jamesburg. Harrison Woodhull Crosby was an assistant steward and chief gardener at Lafayette College in Easton, PA, and he experimented with preserving tomatoes in his Jamesburg home in the 1840s.

In 1847, Lafayette College allowed him to can tomatoes on campus. His process was to solder a lid on a tin can. The lid had a hole in the center that he used to pack the can full. Once full, he took a piece of metal larger than the hole and sealed it by soldering it to the lid. In one year, he packed a thousand cans.

Tomatoes at that time were still a relatively new product to the American public. He couldn't sell them, so being the innovator that he was, he sent them out to people in places of power and newspapers to capture their attention, and hopefully, gain some traction with his relatively new product.

It worked. *The New York Tribune* in 1849 printed a blurb with the title *"Fresh Tomatoes at Christmas"*:

> *Mr. Crosby of Middlesex, NJ, has sent us some fresh tomatoes preserved in tin cans. Whatever the secret of their preparation, we are bound to acknowledge that their preservation has not impaired their flavor. They taste as they would have tasted when plucked from the vine.*

Crosby was recognized by Lafayette College with a bronze plaque on the 70th anniversary of the engineering courses there.

As noted earlier, the North American Phalanx was one of the early canning operations here. One of the larger canning operations started here was the Moorestown Canning Factory. Built in 1868 by Jones Yerkes, it was a two-story operation. Canning was done on the first floor, and wooden crates and boxes were made on the second floor. The can-making operation was across the street. The business grew steadily, and in 1884 it contracted for 500 acres of tomatoes and built a separate facility just for canning tomatoes.

Their intention that year was to can 1.5 million cans of tomatoes. H.K. & F.B. Thurber & Co., a grocery wholesaler and importer, eventually bought the factory from Yerkes, who sold his ketchup through Thurber's grocery stores in NYC. Thurber had their own brand of ketchup, Baldwin Tomato Ketchup, and once they bought the Moorestown Canning Factory, they made this brand of ketchup there.

Moorestown was also the home of the Johnson & Stokes Seed Company, which got its start in 1878. It's interesting to have a seed company and a large canning factory in the same town.

If you look at Moorestown today, you will find defense contractors, premium housing and housing communities, a lot of shopping centers, and the Burlington County Agriculture Center, a connection point for local food.

Canning started to catch on in Salem and Cumberland counties, and by the 1860s the seeds of what would become Campbell's were planted in Camden, NJ. At the same time, J.V. Sharp was establishing himself as a leader in New Jersey's canning industry in Williamstown. Sharp realized that glass was a good alternative to cans for packing tomatoes, and was the first to start using it. Not only did this address a rising concern about the tin and solder used in canning imparting an unpleasant flavor to tomatoes, but the clear glass showcased the bright red tomato color, making for a more appealing product to customers. The canning operation is still in business, and it's the last tomato-packing house in New Jersey.

By the 1870s, Jersey tomatoes were being shipped internationally. The Diamond Packing Company in Bridgeton, NJ, had the capacity to can about a million cans a year of vegetables, sauces, and ketchups. They also packed the "coreless tomatoes" that became well known throughout the United States, and in some parts of Europe.

They weren't the only packing house shipping internationally. The Reverend Isaac W. Dawson opened the first canning factory in Cape May County, NJ, in 1893. Tomatoes were the sole focus of his business, and his capacity was half a million cans yearly. He would ship his tomatoes by schooner to the docks in Philadelphia, where most of his product was loaded directly onto steamers going to England.

While some were shipping internationally, Salem County, NJ, saw a profusion of small canning factories right on farms in the late 1880s and the 1890s. Salem County is a large county, so there were a lot of tomatoes being grown there. A lot of tomatoes are still grown there today. Remember, Heinz was there.

Salem City has a port on the Delaware River, so shipping was available by boat or train. Swedesboro, NJ, was another hub of agriculture business, and there was a railroad line from Salem City to Swedesboro. Farmers found that raising tomatoes for other canners wasn't profitable, so they set up their own operations either in their kitchens or in a barn on their property.

A good example of this was Flora Hancock, who, as a woman taking the lead in the canning business at the time, was extraordinary. She set up her shop in a shed and began to can "frying tomatoes"—basically halved tomatoes that were cored. She became well known for this type of tomato and built a good business, first out of that shed, and then in two locations in Salem City. She closed her business in 1924 when she was eighty.

She was known for her frying tomatoes, but she also canned pumpkin, pears, squash, apples, and berries. While she was successful in her business, once the WWI Armistice was signed in 1918, Hancock, like many canners, had her government contract canceled. Her business got stuck with inventory, and the government didn't pay for the contracted supply. Hancock was able to rebuild, fortunately, and continue to create her place in the canning history of New Jersey.

One last canning company to look at is the South Jersey Canning Company. Freeland G. Sparks and Hamilton G. Pedrick formed a partnership in Pedricktown, NJ, in 1912. There were two operations: one for canning pumpkin, squash, and tomatoes, and the other for saving all the seeds for Pedrick's seed business. The business burned to the ground three years later, and was not rebuilt. Pedrick bought out another canning company that was known for fermented tomato paste, and founded the Alloway Packing Company, which was not for canning but for acquiring seeds.

Whether it was large canning companies like Campbell's, Hurrf, or P.J. Ritter that had interest in seeds and canning, or a small one like the South Jersey Canning Company or the Alloway Packing Company, they didn't know they were forerunners to seed and food sovereignty. They were simply fortunate enough to live that reality as the social norm, instead of having to be advocates and activists fighting for it.

Recipes

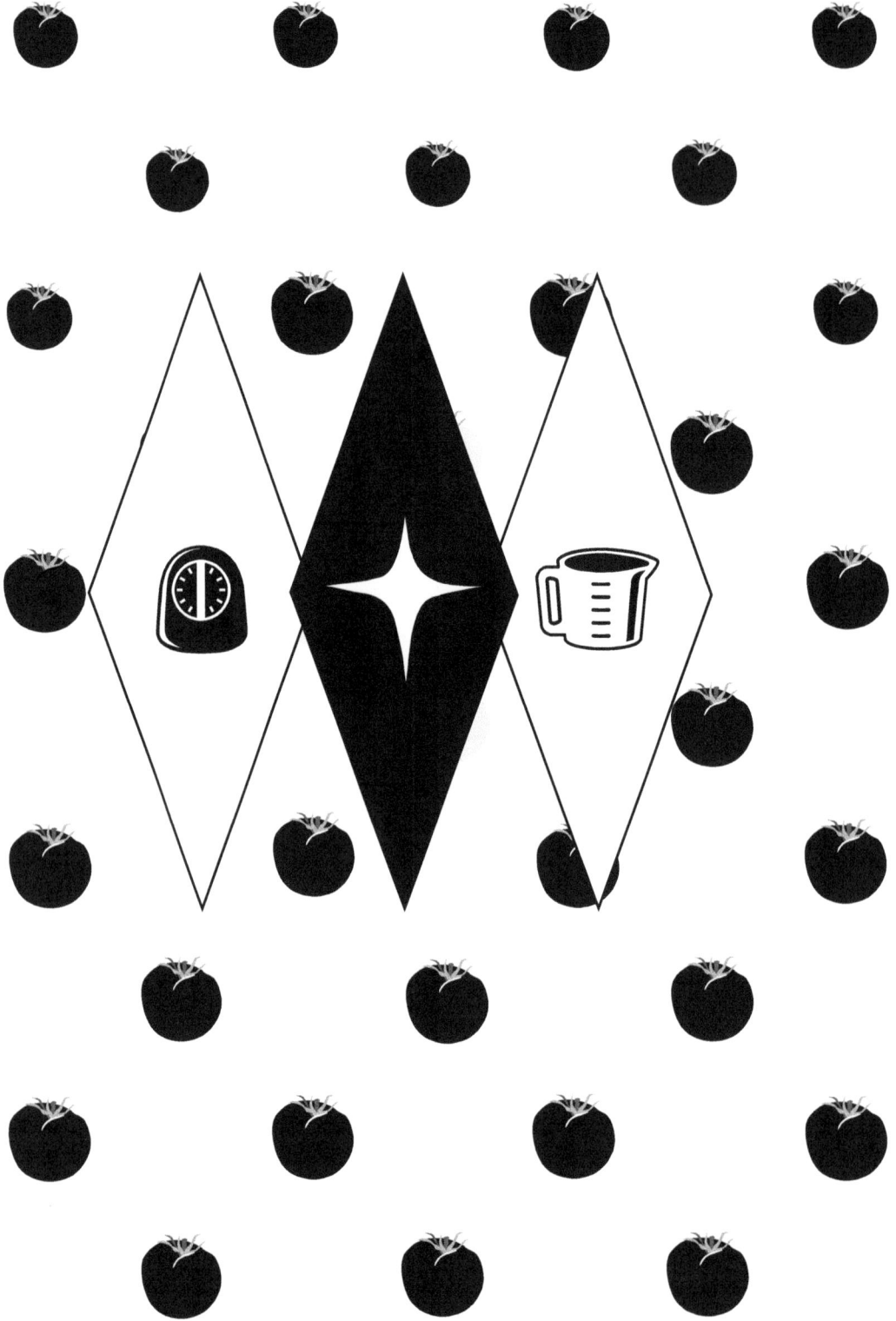

Part of any collective generational experience is recipes.

Family recipes pass down and are often connected to rituals. Local community cookbook fundraisers offered community connection through food. There are traditional cultural recipes and regional ones, too.

An example of a regional recipe is the American culinary tradition of coastal seafood boils: Louisiana Crayfish Boil, Low Country Boil, Chesapeake Bay Boil, and New England Clam Bake.

New Jersey is a coastal state, and we have abundant seafood. So where's the Down Jersey boil?

"Down Jersey" is a local term that refers to the area south of Route 40, which cuts across South Jersey from east to west. It's one of the routes that take you down the shore. The shore in Jersey is down geographically from Philly and NYC. Hence, down the shore. Down Jersey refers to the area south of Route 40 around Cumberland County and the Maurice River, south to the bay, and east to the coast.

Since we have access to the Delaware Bay and the bays and backwaters of the ocean, we crab for Blue Claw crabs.

The closest my family got to a Down Jersey boil was Crabs and Spaghetti. We would catch the crabs early on a weekend morning and make it home by early afternoon to pick tomatoes from our garden, or use what we had on the counter. My mom made crab gravy for dinner. Big bowl of spaghetti with crabs cooked in fresh tomatoes. Boiled, if you will.

The crabs were served on the side and eaten using nutcrackers and pickers to pick out the meat to add to the spaghetti or to enjoy the succulent crab meat by itself. It was belovedly known as "Crab Gravy."

Gravy is a regionalized Italian-American description of tomato sauce. It was gravy for me for a very long time, until I worked with a restaurant family from Naples, Italy. They served traditional Italian cuisine, nothing Americanized. It was in Philly. Innocently enough, one day I mentioned gravy instead of tomato sauce and I got slapped on the back of my head. It was Three Stooges-esque, but with only two stooges: me and the chef.

This chef would always cook a plate for the server who sold the most specials. On many nights, that was me. He was a damn good chef, so selling his specials increased the bill, provided for an excellent dining experience, and meant higher tips. It was a win all around: better tips for me and a freshly made plate for dinner the next time I worked, and he got to show off his mad food skills. We had a friendly relationship. He explained to me that in Italy, gravy was sauce was sauce. This was the first time I had heard of that, so since that exchange, it's been tomato sauce.

Since we would freeze crabs and can tomatoes, Crabs and Spaghetti was always on the table on Christmas Eve for our Feast of the Seven Fishes. I will be sharing that recipe, and a few others that reflect tomatoes and my cultural experiences here in South Jersey, influenced by my Italian heritage.

I'm not a recipe developer. I just know how to cook. Think of these recipes as 3-by-5 inch family recipe cards; a little frayed, but that's because they're beloved and delicious. Simple, fresh ingredients made with love. That is how my mom taught me to cook.

That sets the intention of a *Down Jersey Cookbook*, where the recipes will be developed based on the abundant harvests and prepared with the love and techniques that my mom taught me. Love is cooking and sharing meals. I find sharing meals to be an intimate experience. That's because any family meeting or discussion was around the kitchen table with food.

Tomato Sandwich

Who doesn't love them? The first tomato sandwich of the summer is a cherished ritual of tomato lovers. I make mine as basic as they come: Wonder Bread (toasted), mayo, salt and pepper, and a large beefsteak tomato, preferably a Jersey-bred one.

Sourdough is a good option too, but at least for the first one, it has to be Wonder Bread.

Ingredients
2 slices of bread
1 large beefsteak, tomato
Mayo to taste
Salt and pepper to taste

Directions
1. Toast the bread, or not, you do you.
2. Spread mayo on both slices of bread.
3. Slice the tomato to get the largest slice you can out of it.
4. Place the tomato slice on a slice of your bread.
5. Salt and pepper to taste.
6. Top it with your other slice of bread.
7. Enjoy!

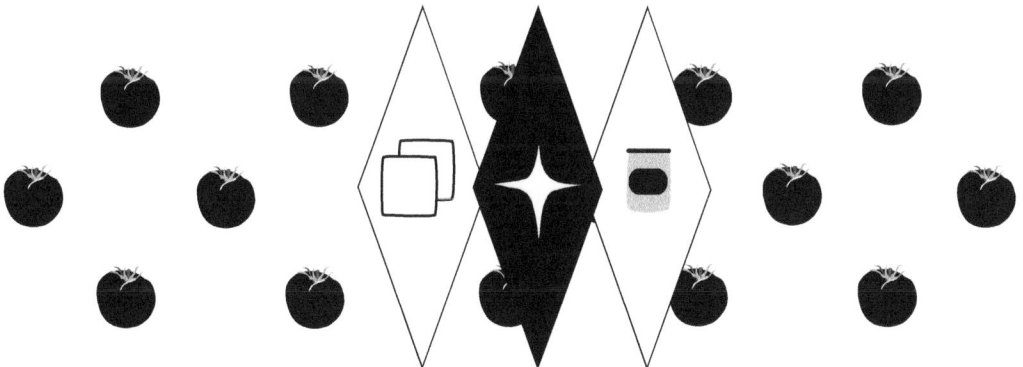

Ketchup

When I read that Edgar Hurff used pineapple vinegar in his ketchup, I was intrigued because I like vinegar. Coconut and sugar cane vinegars are two of my favorites, and I like pineapples. The acidic sweetness of it, I thought, would make an excellent vinegar.

So when you're growing about two miles from where Hurff's canning and seed company was in Swedesboro, NJ, and you're bringing the local tomato history forward with seeds, it's logical to me that you should make some pineapple vinegar. I chose a year, 1903, from his seed lists and grew tomatoes that he was selling seeds for, Sparks Earliana and Greater Baltimore, and enjoyed a taste of Swedesboro history.

So I did. Let me tell you, pineapple vinegar is delicious. I can understand why his ketchup was noted. I used a standard fruit scrap vinegar recipe and just substituted pineapple in place of the scraps.

Since I tried this, I now make small batches of ketchup and store them in jars in the refrigerator. It cuts down on my plastic trash. I don't always use pineapple vinegar; I use frozen garden tomatoes in the winter and when I run out of them, I use crushed tomatoes.

With seeds, seed lists, and pineapple vinegar, I get the basis of a historic local taste that a reputation was built on—and that reputation was built on by a seed guy.

Ketchup

Ingredients

2 lbs. of tomatoes
¼ cup of pineapple vinegar
½ cup cane sugar
1 tsp. cinnamon

2 tsp. vanilla
1 cardamom pod, crushed
Salt to taste

Directions

1. Core and cut the tomatoes up into pieces.
2. Place them in a sauce pot.
3. Add sugar, vinegar, and spices.
4. Stir to mix thoroughly.
5. Simmer on medium heat until thick, adjusting the spices as needed.
6. Let cool, and store in a jar in your refrigerator for a couple of weeks.

Pineapple vinegar

The standard fruit scrap method works fine here, fruit, sugar and water.

Ingredients

1 whole pineapple, cut up with the skin left on
½ cup white cane sugar (it works the best from what I read)
Water
Gallon-size jar

Directions

1. Cut up the pineapple.
2. Add it to a clean gallon jar, leaving about a third of the space at the top.
3. Add sugar.
4. Fill jar until the fruit is just covered with water.
5. Cover the jar opening with cheesecloth, or a tea towel.
6. Let it sit in a cool area away from sunlight for four weeks.
7. Stir it every day for the first week.
8. By the second week, fruit should drop to the bottom of the jar.
9. At the end of four weeks, drain and squeeze all the liquid out of the fruit.
10. If you refrigerate it, it stops the fermentation process.
11. If you don't, it will continue to ferment.

Tomato Soup and Grilled Cheese

We rarely had fresh tomato soup growing up. We were mainly a Campbell's household. All our tomatoes were for sauce, sandwiches, or salads. A can of Campbell's plus a can of milk: we had tomato soup. And it was good.

At that time, tomatoes were still being grown and processed in South Jersey.

This is my recipe for the tomato soup. I freeze whole tomatoes to make it as fresh as possible.

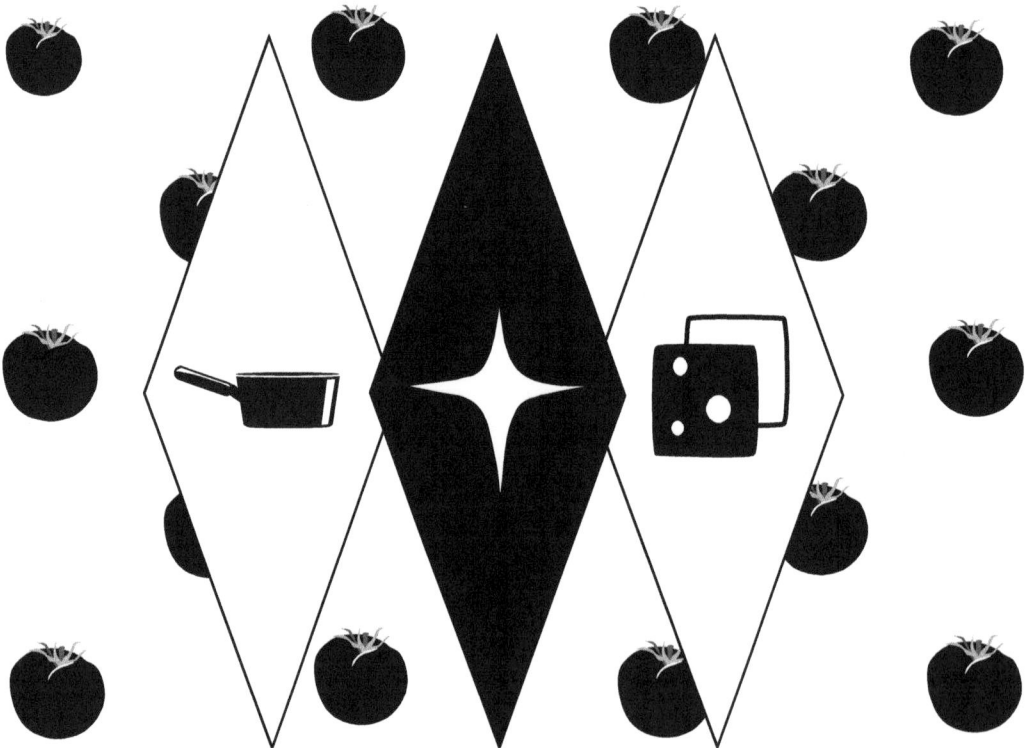

Tomato Soup

Ingredients

1 lb. of tomatoes

3 tsp. of butter

1 large shallot

1 clove of garlic

5 whole peppercorns

1 large bay leaf or two small ones

Salt to taste

Splash of buttermilk, optional

Directions

1. Chop the tomatoes, garlic, and shallot. Rough chop is fine, since you're going to blend it.
2. Melt the butter in a soup pot over low heat.
3. Once melted, add the garlic, shallot, and peppercorns and sauté slowly for three to four minutes, until the shallot is softened.
4. Add the chopped tomatoes and bay leaf, and stir to coat the tomatoes completely.
5. Place lid on the pot, and simmer slowly for 15 minutes, stirring occasionally.
6. Once the time is up, take the pot off the heat and let it cool down with the lid on.
7. Let that residual flavor-induced steam meld into the tomatoes.
8. Once cooled, remove bay leaf and blend.
9. I leave the peppercorns in, but you can remove them.
10. Optional: add just a splash of buttermilk, and stir right before serving

Grilled Cheese

Nothing fancy, just good bread, and sliced cheese. I prefer imported swiss, muenster, and deli provolone. Occasionally Cooper Sharp in place of the muenster.

Ingredients

2 slices of bread

Mayonnaise

Four thin slices of cheese of your choice, room temperature

Directions

1. Heat a skillet over medium-high heat.
2. Place the cheese between the two slices of bread.
3. Thoroughly coat the outside of each slice of bread with mayo.
4. Place the sandwich in the skillet, and cook until both sides are golden brown and the cheese is melted. Serve with soup for dunking.

Meat Sauce

Growing up there was an Americanized Italian meat sauce of ground beef and tomatoes. Ground beef in our house was for meatballs, with pork and veal. Most of the gravy that my mom would cook had meat in it—mostly meatballs and sausage. Occasionally, pork ribs or pork neck bones. Braciole was added for the holidays.

My favorite of all the combinations is slow-cooked pork on the bone in tomatoes with olive oil, garlic, a whole onion, and rosemary. Simmer the pork until you can shred it like pulled pork and serve it on rigatoni, or gemelli, which are two of my favorites for this sauce. Gnocchi is good too.

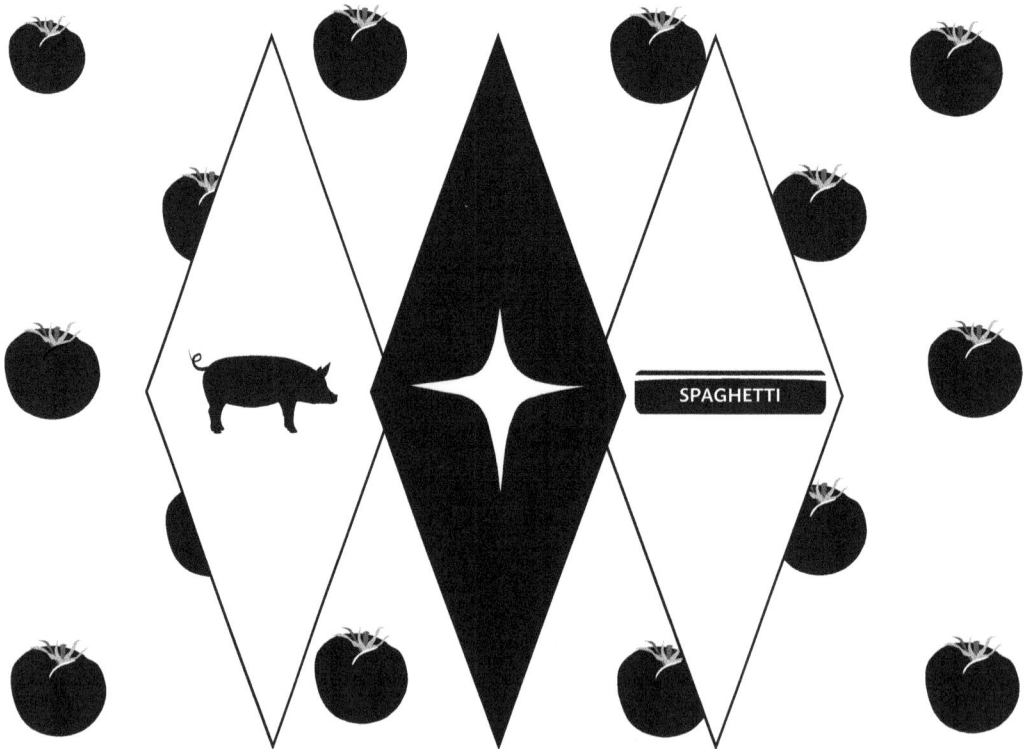

SPAGHETTI

Ingredients

1 lb. of pork, country-style spare ribs on the bone work really well

3 lbs. of tomatoes

Olive oil

Rosemary, fresh or dried

Garlic, minimum of 1 clove sliced or rough-chopped

Salt and pepper to taste

Garlic powder

Directions

1. Coat the pork with salt, pepper, and garlic powder and massage in it.
2. Let it sit for about 20 minutes.
3. While that's resting, put your pot on the stove.
4. Chop the garlic.
5. Cut up the tomatoes, and set aside.
6. When the 20 minutes are up, place a sauce pot on the stove over medium heat.
7. Add the olive the olive oil and heat until fragrant.
8. Add garlic and sauté two to three minutes.
9. Remove the garlic, and set aside.
10. Place the pork in the pot.
11. Brown both sides, two to four minutes on each side.
12. Remove the pork, and add the tomatoes to deglaze the pan.
13. Add the pork back, then the garlic, and rosemary to taste.
14. Simmer for a couple of hours, until pork can be shredded.
15. Shred pork, add back into the sauce, and simmer while cooking the pasta.

Pasta

You can make your own or use boxed. Follow the directions on your package of pasta for cooking. You can finish cooking the pasta in the sauce, or cook it to al dente and toss with sauce in a big pasta bowl.

My mom never finished cooking the pasta in the sauce. It was tossed in a big pasta bowl that it was served in. There's just something about a big bowl of pasta, sometimes being tossed on the table where you're sitting with friends and family that hits different than pasta finished in sauce and nested on the plate. Both are good, just different experiences.

Eggplant and Spaghetti

I love eggplant, but it's a nightshade and out of all of them, this one is the worst trigger for my body. I was diagnosed with Rheumatoid Arthritis and was on a chemo treatment for about four years. My body had a severe allergic reaction to the biological medicine that I was on.

I stopped that treatment, tried one dose of another medicine, and thought, *let me see how long I could go without any treatment*. I'm extremely fortunate; I haven't had to go back. Had the symptoms returned, I would've gone back on some treatment, but again, I'm extremely fortunate. Eggplant reminds me of that if I eat it.

This recipe is from my father's side of the family. My mom's family was all about cutlets, — whether eggplant, chicken, or occasionally veal, — which meant great parmigiana.

It's simply French-fried eggplant strips and a fresh marinara. The key is to be patient when frying the eggplant. You want it to brown and caramelize, but not burnt. Also, you just want to heat it in the sauce, not simmer it. It melts. Add it to the sauce about five minutes before the pasta is done, and stir to coat thoroughly.

Ingredients

1–2 lbs. of tomatoes
1 large purple standard eggplant
1–3 cloves of garlic
Olive oil
Salt

Directions

1. Peel and slice the eggplant in long, thick slices.
2. Cover lightly with salt and let them rest for twenty minutes.
3. Meanwhile, chop the tomatoes, and set aside in a bowl.
4. Slice the garlic, and set aside.
5. After the twenty minutes are up, pat dry the eggplant.
6. Heat some olive oil in a large frying pan and fry the eggplant.
7. Fry until golden brown, place on a paper towel-lined dish.
8. Once frying is complete, place a sauce pot on the stove over medium heat.
9. Add olive oil and garlic.
10. Heat until fragrant.
11. Add tomatoes and simmer while pasta is cooking.
12. Add eggplant back in about five minutes before pasta is done.
13. Once pasta is done, toss and serve

SPAGHETTI

Crabs and Spaghetti

So get this: you boil freshly caught Blue Claw crabs in freshly picked tomatoes and serve it over pasta.

Who knew?

The Italians in South Philly and South Jersey.

While the Delaware Bay isn't the Aegean or Mediterranean Sea, it is home to delicious crabs, oysters, and some fin fish.

Crabs and spaghetti are a traditional Italian-American South Philly/South Jersey dish which aligns it with the American seafood boil tradition. I'm here to say this is a Down Jersey Crab Boil.

Ingredients

A dozen cleaned crabs
2 lbs. fresh tomatoes
Olive oil
Garlic

Directions

1. Mince garlic, set aside.
2. Core and puree tomatoes, set aside.
3. Heat the olive oil in a large soup pot.
4. Add garlic, and sauté until fragrant.
5. Add crabs, stir and place lid on for a few minutes to cook.
6. Add tomatoes, and simmer for 20 minutes. Serve over pasta with the crabs on the side.

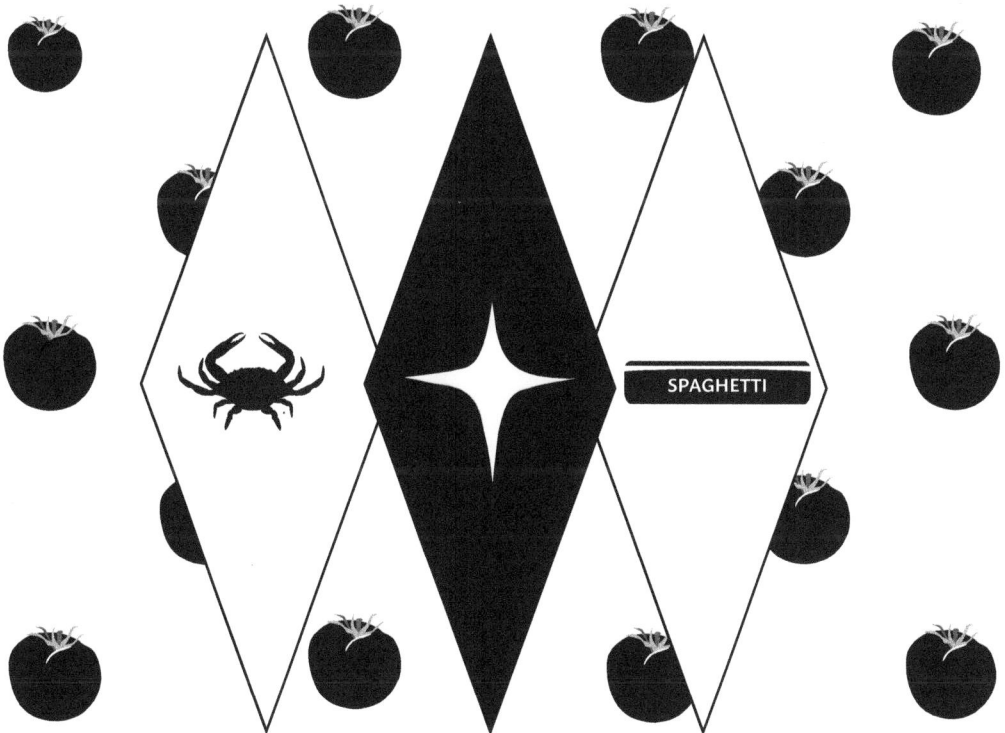

SPAGHETTI

Taking the Extinct
out of
Functionally Extinct

Eye em the Revolution 2025 © Jeff Quattrone 2025

This story of my life started writing itself when my family moved from South Philly to South Jersey. I know it wasn't the intent of my family's move, but just like Mrs. Leeds and the unintended consequence of cursing her thirteenth child, stories are all around us. Her legacy is that her descendants forever bear an association with a beloved local devil.

For me, it's this incredible opportunity to bring forward collective generational stories and experiences around tomato seed breeding that are substantial enough to demonstrate a Jersey way of life.

Including my own.

Who gets to do this? I do, and it's humbling, amazing, and something I don't want to fully believe in. Like all the wonder I felt when I was a kid with the stories on my bedroom bookshelves, or in the third card after the card I was looking for in the card catalog at the local library that I rode my bike to.

I feel like if I do start believing, the story will stop presenting itself —and that's something I don't want to happen.

I'm leading the effort to remove "extinct" from the "functionally extinct" description of Jersey-bred tomato varieties. I want to revive these seeds that were essentially disposed of, since they no longer serve their original breeding function —a true reflection of the disposable society that humans created.

I couldn't do any of this without the seeds and community partners.

These are the Three Rs of iconic Jersey tomato history:

Respect, Revive, and Reinvent.

It's impossible to recreate what was here. What is possible, and will continue, is studying the passion-fueled innovative history and reinventing tomato breeds for where we are now here in Jersey.

The power over the food system is dramatically different now, and so is society. Seeds can help change the existing power dynamic.

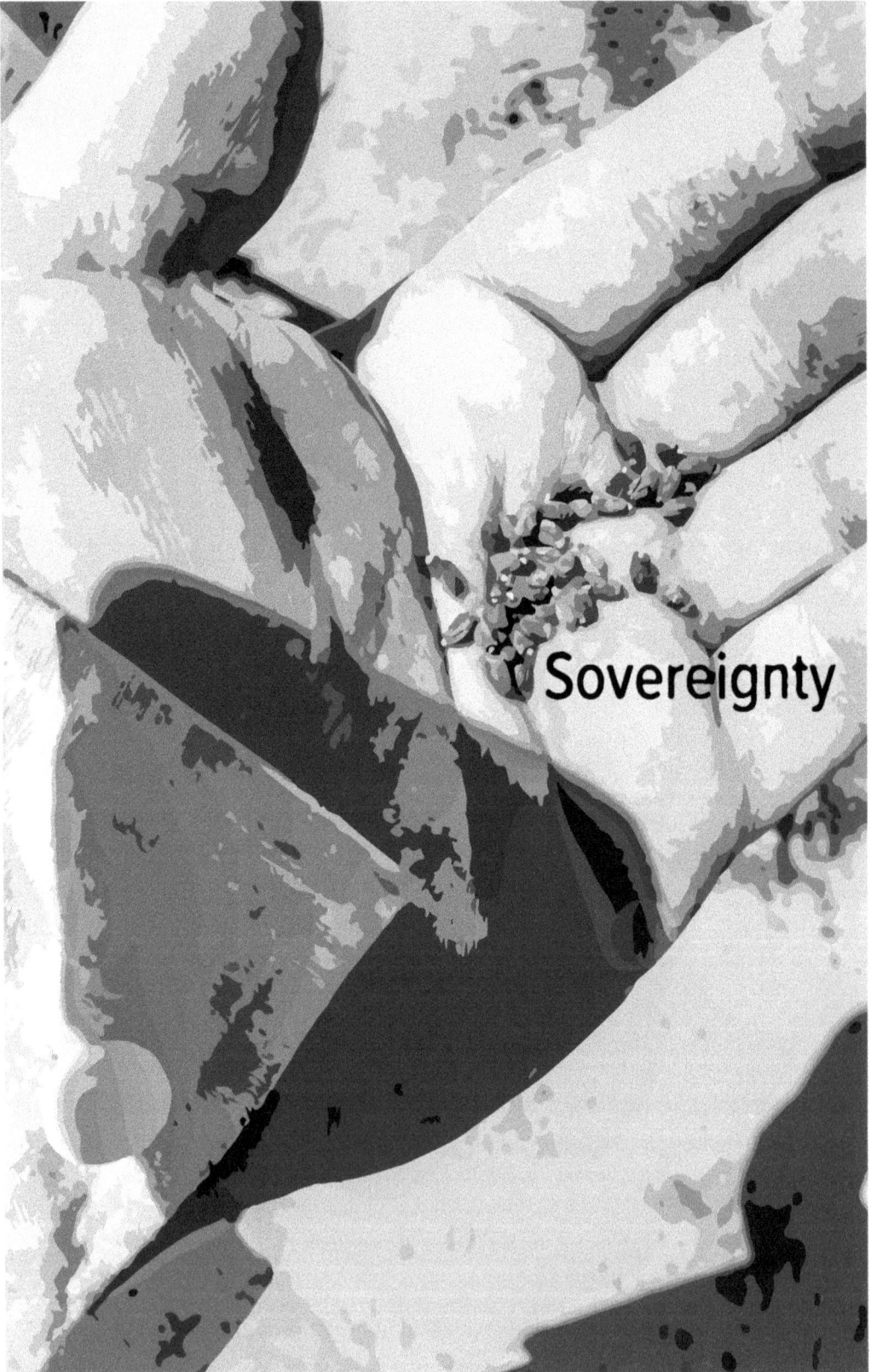

Sovereignty

I sometimes think that what I'm doing is performance art, in that I'm publicly performing what I'm doing by reviving the Burpee Sunnybrook Earliana a couple of miles from where it was selected while reading the actual seed trial journals of that selection process.

One of the takeaways from those budget cuts that led me to art is that art challenges the mores of society. In this case, my performance art is challenging the lack of power of seeds and food by performing local food and agriculture history because I have the seeds and research to do so.

I'm harvesting the revived tomatoes, saving and sharing the seeds, and processing the pulp into trial tomato products with small-scale farmers — at one point working only in a community garden plot, and not as a business entity that is a canning and seed company under one roof with two income streams.

With the price of land in New Jersey, and the lack of canning incubator programs, it's what I can do to take the lessons of history and reinvent them for the time and place of the local food system here.

The seed libraries I've helped create here reflect innovation. They help share the historical innovation of Jersey-tomato seed breeding.

While Campbell's had a closed seed distribution of their innovation, the seeds that were donated to USDA demonstrate the three Rs of Iconic Jersey tomato history.

Respect the innovation and influence of local and national culture. Revive it through grow-outs, seed saving, and sharing through seed libraries. And reinvent it in a modern personal or community way.

And all of this is possible because of seeds.

Index

Acknowledgments

I couldn't do any of this without the help and support of others. For anyone who shared their connection to the work that I do, thank you. When I give a talk, I don't use a slide deck. I have a conversation with the audience. I find it's more engaging that way. I have specific points that I make, but I want to learn from the audience what triggers a memory, a challenge, or a question from my content. It helps me understand the relationship of my work with the public.

People generally don't think of tomatoes as historic, and seed saving as immediate historical preservation. So, the feedback I get helps me understand how to express myself better, and unbeknownst to me, laid the foundation for looking at the collective generational experiences around Jersey tomato seed breeding and the way of life around it. The cultural and sociological connection.

I want to thank my grandparents, Fernado, Maria, Costello, and Josephine. I'm not using their last names because of potential security questions. They left their homeland at a time without global technology that could give them a sense of place of where they were going.

I want to thank my parents, Joseph and Margaret, my brothers Joe and Mark, and my sister-in-law Cathy, for their support. Sharon Furgason and Jess Ferguson were the first two who said yes to seed libraries in Pitman and Woodbury, NJ. Jim O'Connor who supported me from the start and was instrumental in setting up Gloucester County Seed Library in Gloucester County, NJ. It became a model for the county seed libraries in South Jersey. Thanks to the staff at the Cherry Hill, NJ public library who carry on a robust seed library, and Tara Aiken from the Woodbury Public Library who does a wonderful job with the Spring Seed Release every year.

Jessica Floyd for connecting me to the Gloucester County Commissioners for the Kille proclamation. Thank you to the Commissioners also. The family of Willard Bronson Kille for asking if I knew anything about the Kille #7 that started this wonderful, crazy, and fascinating story. Thanks to all the folks who participated in the first Kille #7 grow-out. I lost their names in a computer crash.

Liam and Gabby Duffy from Homebody Farm, who I met through Kathy Simon, thanks Kathy, at the Burlington County Agriculture Center. Liam was instrumental in growing the first crop of Stokesdale, Valiant, and Atlantic Prize tomatoes, and supported the growing of the Kille #7 and Garden State tomatoes. He did this at the Burlington County Agriculture Center, an important local food center. Thanks to Mary Pat Robbie for her vision and commitment to bringing this space to the public.

Laura Simpson for all the support while she was my supervisor. Erin McHugh and Jim McHugh for the friendship, support, and the opportunity for a solo show that resulted in my *Growing the Food Sovereignty Revolution Propaganda Series*. It worked so well with this book. Also to Dianne Rogers, who offered me another solo show of this series.

Mimi Edleman and Mara Welton who I met through Slow Food USA. Mimi is my biodiversity mentor. I'm fortunate to serve with her in leadership of Slow Food USA's Ark of Taste. Mara for her friendship and her keen insight into seeds, food, and the social and cultural impact of it that I get to share. And Kris Reid, thanks for the book launch!

Cookie Till from Reed's Farm, Egg Harbor Township, NJ, which is home to A Meaningful Purpose, a nonprofit committed to community food work, education, and regenerative growing practices. Thank you for your support in my seed and local agricultural history storytelling.

Ed Pappas, Lisa Calvo, and Melissa McGrath from Sweet Amelia's Market and Kitchen who allow me to have a seed garden on their property. It's located less than five miles from where William Maule's Panmure Seed Gardens were. To have the opportunity to do historical seed restoration work a few miles from a former experimental seed garden is rare, and I truly appreciate the opportunity to do this work there in the Pines, listening to Taurus the bullfrog croak away.

Max Sinsheimer for the edits and line notes. He really got to the heart and soul of what I wrote, and when you write about Jersey's heart and soul, one, you better get it right because you don't want Jersey coming for you for anything, let alone about its heart and soul, and two, it's best to have a skilled editor who gets what you want to say, and helps you say it.

Shelly Vitek, thank you for assigning me the article, "The Legacy of Campbell Soup's Tomato Breeding Program" that went viral. Joyce Connolly from Smithsonian Gardens, thank you for your help and for your commitment to the preservation of history. Researching at the Smithsonian on such a specific topic was a highlight of my story. Thanks to the archivist, P.J. Ritter III of the P.J. Ritter Facebook page for sharing so much of Ritter's history. Thanks to Barbara Price from the Gloucester County Historical Society. Bonny Beth Elwell from the Camden County Historical Society, thank you for your help. Lesley Schierenbeck for the StoryCorps interview. Rachel Finn for asking a specific question about including recipes, and all the conversations about this process. Thank you Gail Snyder for our friendship and the beautiful head shot. Mother would be so proud.

Thank you Tish Marks and Kiara Hicks for showing up in a big way early in my community work. I appreciate you both.

Recently, I realized, that I didn't give the Jersey Shore the props that it deserved. Thank you to this magnificent, kitschy, and integral part of Jersey culture. It makes everything all right.

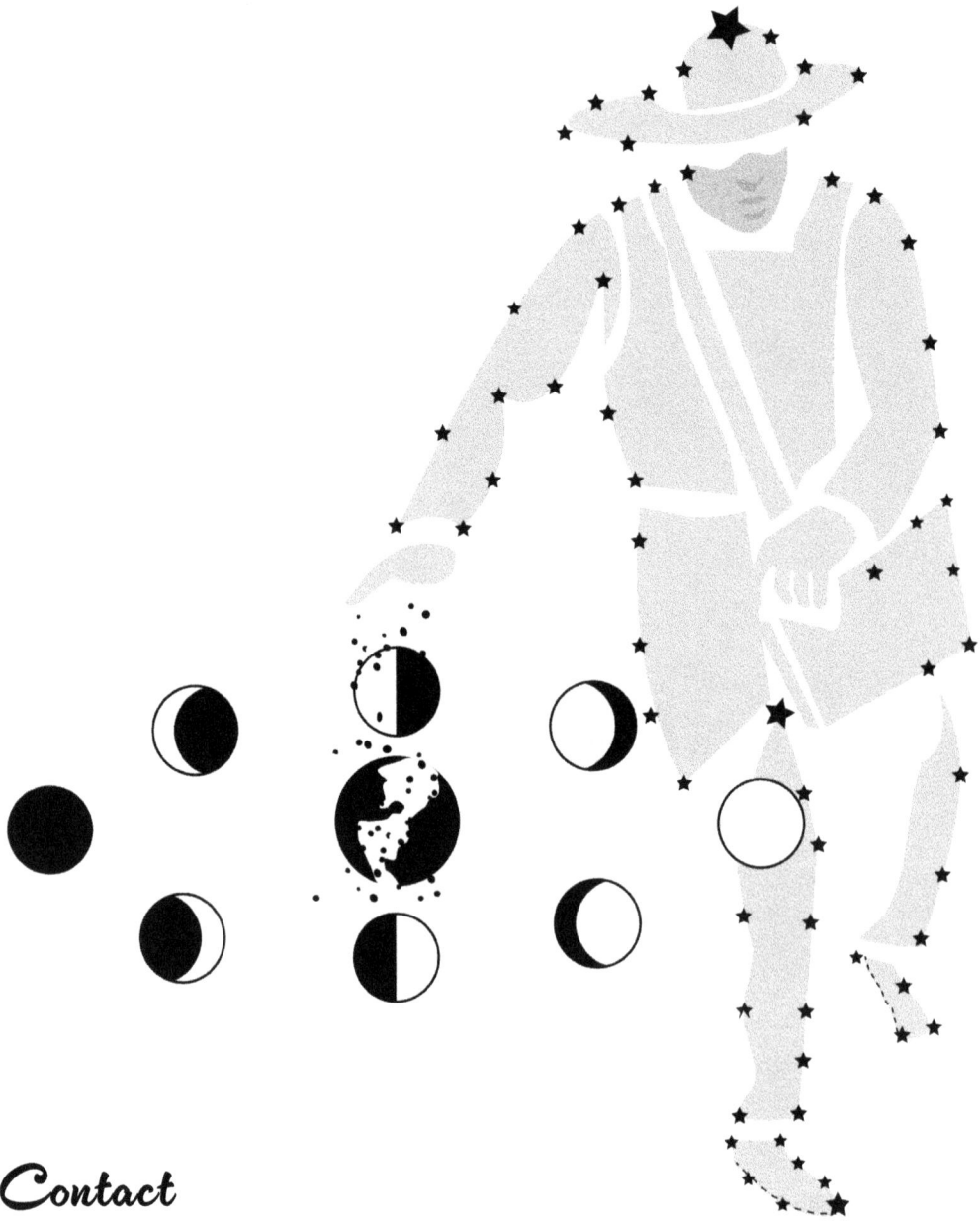

Contact

Iconic Jersey Tomatoes

Art

Moon sowing © Jeff Quattrone 2025

About Jeff

Photo by Gail Snyder
At the Gibbsboro, NJ Community Garden

Jeff Quattrone never expected to be a Jersey tomato seed historian, a leader in the Seed Library movement, and a biodiversity and seed leader with Slow Food USA. Once he found seeds and the stories that they tell, he was drawn into the intersection of the complex relationship between biodiversity, seeds, culture, history, science, and nature. He founded the Library Seed Bank project in 2014, where he brought seed libraries to South Jersey. Once it became a self-sustaining community project, an email popped into his inbox that started his journey into the history of Jersey tomato seed breeding and reviving functionally extinct Jersey tomatoes in New Jersey. His research has been republished by the *Smithsonian Magazine*, archived by Rutgers University, and can be found in the archives at Seed Savers Exchange. His StoryCorps interview is archived in the Library of Congress, and in 2024 he presented a draft version of a seed vision document for Slow Food International at Terra Madre, Slow Food International's global food festival. This seed vision document grew out of a programming decision made by Jeff in 2021 at the launch of the Slow Seed Summit, which that launched Slow Food USA's Slow Seed campaign that he was instrumental in creating. He approaches Jersey tomato seed history by respecting it, reviving it, and reinventing it.

About the Art and Design

All the art and design in this book, including the page layout and both covers, was created by Jeff. Illustrations in this book are modified versions of *Growing the Food Sovereignty Revolution Poster Series* he launched in 2017. Selections from that poster series can be viewed at the Library Seed Bank QR code on the opposite page. This series and other art are available for shows as art, installation, or as historical context. Please use the email QR code on the opposite page to inquire.

www.ingramcontent.com/pod-product-compliance
Lightning Source LLC
Chambersburg PA
CBHW041609260326
41914CB00012B/1432